*Gilles Deleuze's
Philosophy of Time*

Gilles Deleuze's Philosophy of Time

A Critical Introduction and Guide

JAMES WILLIAMS

Edinburgh University Press

© James Williams, 2011

Edinburgh University Press Ltd
22 George Square, Edinburgh

www.euppublishing.com

Typeset in 11/13pt Monotype Baskerville
by Servis Filmsetting Ltd, Stockport, Cheshire

A CIP record for this book is available from the British Library

ISBN 978 0 7486 3853 6 (hardback)
ISBN 978 0 7486 3854 3 (paperback)

The right of James Williams
to be identified as author of this work
has been asserted in accordance with
the Copyright, Designs and Patents Act 1988.

Contents

Acknowledgements	vii
Abbreviations	ix

1	Introduction	1
	Why Gilles Deleuze's philosophy of time?	1
	The do's and don'ts of time travel	7
	The critical power of Deleuze's philosophy of time	16
2	The first synthesis of time	21
	The living present	21
	Synthesis and method in the first synthesis of time	30
	Of pebbles and their habits	38
	The passing present	45
3	The second synthesis of time	51
	A time within which time passes	51
	The deduction of the pure past	59
	Destiny and freedom	68
	How to save all the past for us?	75
4	The third synthesis of time	79
	From Descartes to Kant	79
	Back to Plato	84
	The pure and empty form of time	86
	History, repetition and the symbolic image	95
	Past and present as dimensions of the future	102
	Transcendental dogmatism?	106
5	Time and eternal return	113
	Only difference returns and never the same	113
	Eternal return and death	118

	Series and eternal return	124
	Disparity and difference in eternal return	130
6	Time in *Logic of Sense*	134
	Of wounds and time	134
	How much, how and where?	138
	Time and the surface between depths and heights	145
	From principles to acts	154
7	Conclusion: the place of film in Deleuze's philosophy of time	159

Endnotes 165
Bibliography 194
Index 203

Acknowledgements

I am grateful to the Carnegie Trust for the Universities Scotland for support for research on Gilles Deleuze in Paris and Bordeaux. Dundee University supported this book through research leave and support for my postgraduates through successive PhD scholarship schemes, alongside grants from the Arts and Humanities Research Council and the Overseas Research Students Award Scheme. The School of Humanities provided conference travel grants for many Deleuze conferences and workshops. Once again, this work benefited greatly from research with my current and past PhD students working on Deleuze and French philosophy across many fields (Paul Barlow, Yannis Chatzantonis, Tim Flanagan, Carrie Giunta, Jenny Kermally, Andrew Mcdonald, Neil Mcginness, Stefanos Pavlakis, Aude Pichon, Fabio Presutti, Brian Smith and Dominic Smith). The projects run by the students on my philosophy of time and Deleuze modules at Dundee have greatly influenced this work; I am grateful to all of them. I taught at successive Deleuze camps at Cardiff and Cologne, as well as at the Melbourne School of Continental Philosophy, in the years preparing for this book. The intensity of enquiry and the desire to share discoveries among all teachers and participants at these events shaped my thinking and renewed my enthusiasm for Deleuze, allowing this book to be much better than it might have been. Of course, if it is still much worse than others might reasonably expect, that is entirely my fault. There are many other blameless individuals whose thoughts, works, conversations and critical comments helped this work; here are but some of them: Jack Reynolds, John Mullarkey, Beth Lord, Keith Ansell Pearson, Rachel Jones, Patricia Pisters, Roger Young, Guy

Callan, Lily Forrester, Jon Roffe, Jeff Bell, Ian Buchanan, Nicholas Davey, Nicholas Blincoe, Johanna Oksala, Mike Wheeler, Levi Bryant, John Protevi, Giuseppe Bianco, Andrew Benjamin, Miguel de Beistegui, Dan Smith, Amanda Montgomery, John Drummond, Carol Macdonald. Finally, and not only because you dared me, *this one's also for you love* . . .

Abbreviations

In the text, references to *Difference and Repetition* are given in the form (DRf, page reference) for the French edition and (DRe, page reference) for the English translation. References to *Logic of Sense* are given as (LoSf, page reference) for the French edition and (LoSe, page reference) for the English edition. Many translations are my own or are substantially modified so there are more references to the French editions; however, there are frequent cross-references to the English editions in order to allow for easy tracking and checking of the texts.

For students past, present and future,
because they revolt and disbelieve . . .

1

Introduction

WHY GILLES DELEUZE'S PHILOSOPHY OF TIME?

Why study Deleuze's philosophy of time rather than any other aspect of his philosophy? Why give time such prominence among the rich and varied lines of thought he gives to us? Two answers guide the following reading of the work on time in Deleuze's *Difference and Repetition* and *Logic of Sense*. First, in those books, Deleuze sets out one of the most original and sophisticated philosophies of time to have appeared in the history of philosophy. It ranks alongside the work on time achieved by Kant in the *Critique of Pure Reason*, Heidegger in *Being and Time* and later essays, and by Henri Bergson in *Time and Free Will* and *Matter and Memory*. It allows for productive debates with the emerging scientific and philosophical views of time in twentieth-century physics, not only in terms of dynamic systems and relativity but also in terms of quantum mechanics. It closes but also interacts with work in phenomenology and post-Kantian transcendental philosophy. Deleuze's philosophy takes seriously and even extends Nietzsche's extraordinary suggestions about time and eternal return. It responds to Platonic theories of circular time and other theories of metempsychosis. It renews the philosophy of time as studied by the Stoics. The philosophy is partly developed in relation to theories of time as they relate to modern theories of the unconscious. Finally, and perhaps most importantly, it suggests a metaphysical model of time relating to problems concerning time in evolutionary theory and in questions of genesis as they impinge on action. In short, Deleuze's philosophy of time is a new philosophy of time at the crossroads of the history of philosophy, of modern sciences and of contemporary problems about time. As such, it cannot be ignored whenever the philosophy of time is to be considered today, even if such consideration is eventually to reject Deleuze's speculative and metaphysical approach in favour of other sources.

This influence of Deleuze's philosophy of time is behind my decision to have a comprehensive set of endnotes in this book. The study of Deleuze's work is now a mature and perhaps even somewhat crowded research area. There have been many excellent recent works on Deleuze and nearly all of these books and articles touch on Deleuze's philosophy of time; however, the way they do so is varied according to the methods and topics selected for the interpretation of Deleuze's philosophy. One of the purposes of the notes is therefore to provide a range of alternative and sometimes cognate positions to my own reading of his philosophy as a process philosophy of time. No doubt this angle has been influenced by a reading of Whitehead that has been deliberately kept in the background here, lest it skew the work on Deleuze.[1] It is hoped that the notes will allow readers to set their own paths through a body of work that has now been carefully charted by many dedicated researchers from many different subjects, but that remains extraordinarily rich and responsive to variations according which methods are ascribed to Deleuze's philosophy, which of its roots are taken to be the most important, and which of its concepts are given the starring roles. The beauty of his philosophy of time is that it gathers all of these problems and questions at their most urgent yet also most rewarding knots.

The second reason for giving prominence to Deleuze's philosophy of time is that it is at the heart of his early philosophy, the period before the joint works with Félix Guattari. It is still important after that, notably through the role of history in their works and in Deleuze's works on cinema (to be discussed in my conclusion), but its role is more peripheral and, arguably, less significant than the operations taken up by spatial concepts. We shall see that the crucial role of time in the 1968 and 1969 works rests on Deleuze's work on serial processes, on repetition for itself and on difference in itself. The innovations on time allow him to explain and develop a philosophy of process and becoming without having either to ground it on a prior foundation dependent on some kind of metaphysical identity, or to give it some kind of independent experiential or empirical basis, or to make mystical and quasi-religious claims for its legitimacy and form. This philosophy of time is therefore one of difference as becoming, where difference is free of any roots in identity claims or in analogies based on sameness and similarity. The new theories of time constructed by Deleuze allow difference to be defined as a complex process the detail of which can be explained through the claim that time itself is process.

Yet this first step on difference and process is itself insufficient for grasping the originality of Deleuze's ideas, since the claim is not strictly about time as process, but rather about processes making multiple times. Times are made in multiple synthetic processes. Later in this book, when we encounter his famous three syntheses of time, we shall learn to read the syntheses as the dominant term in the conjunction: time is the result of the syntheses and not the other way round. These multiple times cannot be reduced to one another according to an order of priority or even an overarching theory of relativity borrowed from physics. Instead, the processes and the times are related according to a series of irreducible perspectives, if we are to use a metaphor. Or they form a network of asymmetrical formal and singular processes, if we are to be literal and technical.

What is a network of asymmetrical formal and singular processes? It is a network of related processes operating on one another but in different ways depending on which process takes another within itself. So for example, we shall see in the study of Deleuze's first synthesis of time, in Chapter 2 of this book, that the synthesis implies a process in the present determining the past and future as dimensions of the process. This already determines two different processes: the contraction of the past and of the future into the present. In Chapter 3, however, we shall see that if the synthesis of the past is taken as the primary process, then the present becomes a dimension of the past, as its most contracted leading tip when we picture the past as an expanding cone. This in turn adds another process to the two already defined in relation to the first synthesis. Most significantly, it also means that the present is not one process but many, dependent on its place as primary synthesis or as dimension. So Deleuze is giving us a formal network of processes defining time as multiple, because any reduction of this multiplicity is disallowed in Deleuze's construction. We shall see that this concept of multiplicity is crucial to any understanding of Deleuze's philosophy of time. There is no theoretical way of treating each of the different relations as if they conform to a single overarching law, logic or process. When interpretations of his work impose a further identity on this multiplicity, such as the idea of multiplicity as One or Creation, then they misrepresent his philosophy of time.

The main reason there can be no such reduction lies in Deleuze's use of asymmetry, a very important term in his philosophy of time, since it not only explains why time cannot be reversed and why there is an arrow of time (or rather a series of arrows depending on which process we take), but it also explains why processes themselves cannot be reversed. The counter or reverse of any given process does not go back to an original position or state, preserved according to some set of laws or kind of symmetry according to isomorphic functions, but rather it brings about another transformation that is itself irreversible. So the transformations implied by Deleuze's multiple view of time are all irreversible and asymmetrical. There is no going back because the initial conditions have been changed by the process such that even if we were to reproduce, for instance, an initial set of objects, the place and function of those objects within the processes will have changed. In turn, this is a first clue as to the radical nature of Deleuze's philosophy of time: it is inherently anti-conservative and anti-reactionary due to its inbuilt and unavoidable asymmetries of time. There is no represented and original past to go back to. There is no eternal realm to escape to in the future, where time stands still. Every process is multiple, irreducible to others and free of claims to higher sources or pure origins. There is no way back and no way up and out. There is only the demand to be worthy of the complex processes making all things become together, but never as one.

Why, though, have I added the term 'singular' to the description of Deleuze's formal account of a multiplicity of times? It is because the way time is made according to a network of processes is essentially singular. There are different times according to the singularities of the individual processes at work. We could assume that the multiplicity of times described by Deleuze is constructed out of many different times, each one of which provides a setting, or field, or enclosure for perspectives on objects, or subjects, or relations, or individuals. They would occur in those many times. This would be wrong, however, because he is reversing the familiar commonsense order of time and of those things assumed to occur in it. For Deleuze, processes make times and those processes are determined by singularities rather than by regular features adapted to general laws and relations. So though he speaks of formal relations, these are not fully determining of times. The formal frame describes types of processes, for example, those determining the past as a

Introduction

dimension of the present; however, within that frame singular events determine their own times. We must therefore speak of many presents with their own ways of taking the past and the future as dimensions. We must also avoid any general spatial representation of time as something pre-existent that things can be placed on or in. There is no general line of time and no space–time continuum. Instead, singular processes make their own times within the limits set by some wider formal principles, such as asymmetry.

Time for Deleuze is therefore not only irreducibly multiple at the level of types of time: present, present as dimension of the past, future as dimension of the present, future, and so on. It is also irreducibly complex insofar as each one of those types can only be said to be fully given when it is associated with singular events, which are themselves determined in accordance with series of singularities or singular processes drawing events together as processes of becoming that make times. We shall see through this study that this multiplicity and this complexity raise serious critical objections concerning the determination of times and the explanation of whether and how there can be relations across singularly determined times. First, there are many problems concerning the communication and inter-relation of different times. If one event determines times singularly in a different way from another, how can those events be said to be related without setting an overarching time or theory of the relativity of time accounting for that relation? Second, if time is to be determined by events, there are formal problems concerning the number and scale of times and events. Does any event and singular process determine times? In which case, do we have to assume infinity of times divisible according to smaller and smaller events? Which should then be given priority to, if any? Third, this focus on events and singularities risks reintroducing a subjective or human centre to time, whereby times would be recognised and determined according to experiential and potentially subjective assessments of what counts as an event and as singular. Of course, some might want time to be human-centred, but this is not the aim of Deleuze's philosophy.

The implications and difficulties of Deleuze's position can be drawn out thanks to an example. This example simplifies things and is no substitute for the full descriptions to follow in the main body of this book. Its goal is to give a basic explanation. Every day, in a corner of a Victorian cemetery, a few feet away from the grave marking the death of a loved one, an ageing figure copes with grief and loss by feeding pigeons and squirrels at the foot of a decaying statue. None of these actors and things exists in the same time, the same present. The events around each singular thing, the events whose synthesis creates the thing, determine a present. This present includes other things, but its singular synthesis does not include their present in the same state or in the same way that they include it. There is an asymmetry between the times despite their reciprocal inclusions. The hand repeatedly spins nuts and crumbs out into the air, distracting the injured brain and unconscious from an obsessive return to a devastating absence. In so doing, it draws the past and the future into its present struggle according to the event of moving on from a death that threatens to drag other lives down with it. The predatory fight between pigeons and squirrels for dominance of the sanctuary defines a different present, a longer one, in the sense of the extension of the waves of its contractions over generations and over space according to the fluctuation of populations in line with food supplies and disease.

The presents include one another, but they do not do so without transforming each other, dependent on which event is set as the wider inclusion. There is a present where the pigeons are peripheral to a mental and physical resistance. There is another present where that resistance is of minute and contingent significance in much wider fluctuations of animal populations and evolution. There is hence an asymmetry between the grief-event when it is at the centre of its own synthesis of time, and its capture in the present of the fight for survival between animal populations. Each process therefore determines its own times, for instance where the future as dimension for present grief is very different from the future as dimension of survival of an incoming species and extinction of another. Similarly, the past threatening to engulf a grief-stricken lover, where the present becomes a dimension of that grief, is different from the past engulfing a fleeting skirmish between a bird and rodent. The future determining a present as its dimension, through the flicker of a new interest in life raised by the repeated act of feeding, is not the same future determining the perishing of the last red squirrel in the cemetery as another species outfights it for the new source of food. All these times are relative to one another, not as disinterested objects, but as processes folded into one another yet incompatible with a representation of all of them under a single external map or in a single space–time continuum.

The difficulties for a multiple philosophy of time, even in this simple example, can be seen in the contingency of the selection of each event. Why focus on the grief and the human agent? Why on

Introduction

the animals rather than the trees? Why not subdivide each tree into leaf-events, each with its own present, then leaf-vein events, then events on tiny portions of blades, then on even smaller patches between those blades? What are the principles guiding these selections and, in the absence of any such principles, can we not divide time according to an infinite spatial subdivision of a contingently selected portion of the cemetery and its events? It is even possible – indeed I will argue that it is necessary – to determine multiple times not only in terms of living beings as agents and as members of evolutionary lines. The stone statue with its rapid erasure of distinctive features due to the fragility of the local limestone also determines a present and a synthesis not only of its past, but of the droppings of the birds, the breath of passers-by as they come close to read an inscription, the damaging variations in light, humidity and temperature. The eroding stone marks a present contemplating the passing of the human and animal lives around it. Even a judgement of damage is relative to different times, to whether they are dimensions of a new event or of the passing of a living present. The challenges for an account of Deleuze's philosophy of time are therefore great, not only in presenting its originality, but even more so in the fragmentation implied by its multiplicity and in its retention, despite this division, of a sense of communication between irreducibly different times. To give a more intuitive sense of both angles, I will now give a further example designed, first, to give an intuitive sense of the implications of the philosophy of time, then second, to show its potential despite or perhaps thanks to its complexity.

THE DO'S AND DON'TS OF TIME TRAVEL

Philosophies of time and metaphysics impose a variety of restrictions on the possibility and form of time travel. Exactly how this occurs is a good intuitive way of grasping the manner in which apparently abstract commitments with respect to time have repercussions on what can and cannot be envisaged in a practical context. A further instructive consequence of these intuitions about practical consequences can be found in the way they also reveal the a priori, pre-empirical commitments of a philosophy and of its methods of construction and deduction. These revelations can be devastating for a philosophy, for instance, where commitments incompatible with empirical facts are brought into the open. This is particularly relevant to work on Deleuze's philosophy as it is to be studied in this book, since its methods will be shown to be distant from scientific empiricism and to involve bold speculative and transcendental moves inviting critique not only on the basis of current (and indeed of outdated) scientific positions, but also from the broader point of view of empirical scepticism.

If we begin from a very general definition of time travel as moving backwards into the past and as skipping forward in time, four general features of Deleuze's philosophy of time can be drawn out. According to his philosophy of time any process in the present is also, in some special way, a process in the past, a moving backwards into the past. Any process in the present is also, in some special way, a skipping into the future. However, no process in the present can go back to the past as it was when the process of going back began. No process in the present can go into the future as it will be when the process of going into the future began. I have separated travel back and travel forward in time here because the 'special ways' indicated are very different. It is important to stress that the time travel considered is actual and not merely virtual, ideal or imaginary. Deleuze's philosophy allows for actual time travel back and forward in time and this depends upon virtual time travel, in a very special sense of virtual, for some but not all of its features. In fact, it does not only allow for time travel, it makes it inescapable. The past and the future are not simply realms we might be able to visit, they are processes fully implicated in our present ones. They are not only present as mediated, in imagination and memory, but directly in all present processes.

So Deleuze's philosophy does not allow for time travel as it is popularly conceived, for instance in science fiction, because any travel that can take place presupposes a transformation of the past and of the future. So even if we could use a machine or property of relativity or quantum mechanics to travel back or forward in time, the time we would arrive at would not be the time we aimed for as 'the past as it was' or 'the future as it will be'. You never go back to a past that preceded you and that you might want to correct. You never move forward, jumping over the future determined by the present. So you can never jump to a point where you might want to tweak those pre-determined patterns. Instead, and perhaps even more surprisingly, according to Deleuze's account we are travelling back and forward in time all the time and with no need for special machines or for odd physical properties such as wormholes. This is because the 'special ways' included in the statements above involve alterations of the past and of the future in the present. Any actual present process is altering the past and the future, not in a causal or statistical manner but instantaneously and for all the past and all the future. So as you read this you are a time traveller, as defined above, but then so is the cat at your feet and, for that matter, so is the wooden bench you are sat on. On the other hand, the slightly tousled astronauts burning holes in your floor with their smoking boots have not travelled back to their past and they cannot go back to your future either.

The best way of understanding these statements is through Deleuze's idea of times as dimensions of one another. For him, past, present and future are not separate parts of time. Instead, they alternately treat each other as dimensions, where to be a dimension means to be a subsequent process. These processes operate on series of events. Dimensions are therefore decided according to prior and consequent processes operating on series. There are prior processes according to which others are only dimensions, because the dimension follows from the prior process rather than having an independent status of its own. So Deleuze's process philosophy of time denies the independence of past, present and future. Instead, it sets each one into many different orders of dimension according to many different processes. These are sometimes organised according to the traditional concepts of past, present and future, but not necessarily so. For example, in *Logic of Sense*, the first order organisation is between two times taken from the Stoics: Chronos and Aiôn. As we shall see in Chapter 6 below, this order is then subdivided through all the processes operating with each time and, more importantly, across them.

It is important to register that although we can speak of different processes and orders of priority, this subdivision does not imply that different orders of dimension and processes are completely separate from one another. It is exactly the opposite. The processes interact and the orders include one another. However, they do not do so according to a final reductive unity. The picture corresponding to this would then not be a table of separate processes or a single set of laws but rather a many-dimensioned web with many different kinds of process operating in it and interacting with one another. Gusts of wind pull the web and give it concave and convex shape. Damp and raindrops weigh it down. The spider and trapped fly set off and respond to intricate vibrations. The strands wear and decay and are renewed in patterns of varying strength and robustness. Each process can be taken as the prior one, for instance when the wind and the damp are taken as factors in the prior struggle of the fly to escape before the spider reaches it, or when we blow on the web to make a certain shape and hence view the trapped fly as an impediment to the search. Against this view of a complex and irreducible multiplicity of dimensions, there is the temptation to arrive at a final representation and model for all the processes together, such that a single view and theory is sufficient to account for all the interactions according, for example, to a single set of laws. It is this temptation and possibility that Deleuze's philosophy of time seeks to resist. One of the most important critical questions will be whether he is successful in developing a genuinely multiple philosophy of time, rather than a heterogeneous image open to a later reduction into a homogeneous whole, or to an atomisation into completely separate processes.

In *Difference and Repetition*, time is defined through three syntheses, where each of the three syntheses is a prior process in relation to the other times as dimensions. For instance, as we have already noted, in the first synthesis the past and the future are dimensions of the present. This means that a process in the present, often described by Deleuze as the 'living present', determines processes in the past and in the future such that the present process changes the past and the future by synthesising them in a novel manner. The particular form of this synthesis is contraction, in the case of the first synthesis of time. A mistaken understanding of these syntheses points to the correct one. When you stupidly leak out a secret to someone who really should not have been told it, you might well view this as a betrayal of the past and as an act setting off an undesirable causal chain of events in the future. As such, you change the outcomes of the past and set off a new causal chain into the future. This is not what Deleuze's synthesis implies as contractions of the past and of the future. Instead, in the first synthesis, past events and future possibilities become fully actual, fully present. So all of the past and all of the future are concentrated into the living present, into the multiple syntheses of living presents. Your incautious slip does not change the outcomes of the past but the past itself, which has no existence independent of its contraction in the present. The slip does not set off a causal chain. It contracts all of the possible chains in the future in a different way.

In Deleuze's first synthesis of time, past and future are therefore dimensions of the present such that each present is infinitely extended back and forward in time. Time then becomes a set of many different and ever wider presents, where wider does not mean less limited back and forward in time, since they are all infinite. Instead, wider indicates which and how many chains of events are

Introduction

taken as prior. These include one another like interlocking containers. However, because each living present determines a series of such relations, we do not have one order of containment but many, with each one determined by a living present and a subsequent order of containments. The present of the cat at my feet is contained by my living present. This means that its past and future are determined as dimensions, not by its own present, but by mine. However, there is also the cat's living present for which my future and past are dimensions. So when I forget to feed the cat, there is an order of inclusion where the cat's hungry future is unimportant. The cat moves from my side and meets some ninth misfortune in search of sustenance, but this is a minor event compared with the struggle to finish a few pages on the philosophy of time. There is also, though, a living present for the cat with a future dimension concentrated back on to its hunger and my abject failure of care, where my future dimension as unreliable writer is utterly trifling. All of these times interact and interlock, but according to Deleuze's philosophy of time they do not submit to any external or internal order that can reduce the multiplicity of times to a single set of laws, patterns or even probabilities.

The first synthesis of time, the living present, determines each present as a time traveller not only unfolding out into the past and future it concentrates into itself, but also folding into the living presents containing and transforming it. We do not travel through time as an unchanging body and mind. Instead, all present processes resonate in waves through the past and future as dimensions of events determining a living present. Metaphorically, the first synthesis of the living present is a series of patterns of interference concentrating onto a condensed core of singular disturbances, like drumbeats on skin covered with coloured patches of sand. One side is the past and another the future, each shaking and forming different shapes as the drumsticks hit towards the centre. Except that, like so many of our most accessible examples and metaphors, this draws us too quickly to a sense of conscious action at the centre, in the present, whereas we should imagine a feedback loop from the patterns to the arms driving the sticks, such that each beat is driven passively. Each beat is forced to improvise according to unconscious impulses, by the pattern in multiple loops that allow for decision, selections and choices, but only as partial overlay upon many more passive processes.

Yet these waves concentrating past and future into multiple interlocking presents are but one side of time. The second synthesis of time from *Difference and Repetition* also determines the other times as its dimensions. Along with Deleuze's take on Nietzsche's doctrine of eternal return, this synthesis is perhaps the one furthest from common intuitions about time. Such intuitions are not necessarily a good guide for consistent and adequate models of time. They do however present a barrier to communicating philosophies of time distant from common experiences, even if such experiences are themselves formed by language, context, history, concepts, bodily similarities and differences, and presuppositions about sense and nonsense. Two settings are helpful for an understanding of the second synthesis or the pure past, to be covered in depth in Chapter 3 of this book. First, the second synthesis runs counter to the notion that the past is what we can represent as past, not only in conscious memories, but also in descriptions, archives and media such as film and photography. This past is pure, in the sense that it does not contain entities open to representation. Second, the second synthesis is a process operating on the present, which becomes a dimension of the past. The pure past makes the present pass.

When considered together these settings are a formidable challenge. Deleuze has to construct an account of the past that is: (1) not a representation of formerly present events; (2) resistant to later representations or in excess of them in some way; (3) works on the present and makes it passive in relation to the past, to the point where we can say that the past makes the present pass away into it. Frequently, explanations of this philosophy appeal to its roots in Deleuze's interpretation of Bergson. In Chapters 2 and 3 – indeed, throughout this book – I will make constant references to the influence of Deleuze's studies of other philosophers on his philosophy of time. Nonetheless, here, I shall give a simplified version without such background references and assumptions. This is partly because I regard them as restrictions of scope in audience, for instance if the reader has no experience of Bergson. It is also (and probably more controversially) because such appeals can sometimes create a form of obfuscation. This dents critical and original understanding by replacing robust, rigorous and inventive examination of concepts and arguments with appeals to authority, whether of the history of philosophy itself or of the holder of historical knowledge. Finally, and in a related fashion, it is because Deleuze's philosophy of time is as important for its originality, for the way it departs from the positions it also draws on. When it is explained on the grounds of Bergson's philosophy, or Kant's, or Hume's, or Nietzsche's there is a risk of blunting originality and falling

Introduction

into stultifying questions about fidelity to original intentions or texts.

If the past is to avoid representations as content and the ability to be represented as formal condition, the past will have to be pure and resistant in some way. It will have to be free of representations and capable of undermining them when they are imposed upon it. If the past it to make the present pass, it will have to include it in some way. This is why the present becomes a dimension of the pure past, where the past is a series of levels of pure differences. Each of these levels comprises all the relations of pure differences as changed by a passing present. Two terms require a lot more explanation here: pure difference and passing present. A pure difference in Deleuze's philosophy can be understood as an abstract idea associated with a becoming if we follow *Difference and Repetition* (to become weaker, say), or as an infinitive if we follow *Logic of Sense* (to weaken). Deleuze's argument is that in order to be a living present, that is one in a state of change, the present must express or actualise a multiplicity of pure differences. So, for instance, when a thing loses its power to do something, it is expressing or actualising the idea of weakening in a singular manner. As such, it is altering the relations of the idea to all others (or of the infinitive to all others). This 'all' is the level of the pure past expressed in the present. It makes the present pass because it sets it as the limit of all pure differences making it become other than what it is.

A good way of thinking about this is through Deleuze's concept of dramatisation and the example of an actor. We are all actors for Deleuze, replaying and relayed by the pure past in novel dramatisations. Our challenge is to be worthy of the pure past, of its multiple relations of pure differences. So when actors replay a moment of humiliation, for instance, they are not replaying any particular representation of it (*I'll do it exactly as Brando did it*) but are instead trying to express humiliation in a singular and new circumstance (*how can I make it work with my body, with this audience, after these events, for these people, tonight?*). But even as they enact the affect, the singular events are passing and fading away (*the decompression even in the joy of the curtain call*). They pass away exactly because any representation or repetition of them in the same way fails to capture the first singularity (*You had to be there!*). So what drives this passing is also what makes the representation fail. It is the relation of pure difference, the essence of humiliation, to its singular expression. It is the pure difference (the idea of humiliation, to be humiliated) that makes the present pass. This passing is not though restricted only to humiliation. It depends on an expression of humiliation in its relations to all other ideas and infinitives. When the actors express humiliation in their singular manner they change its relations as pure humiliation to courage, despair, strength, loneliness, compassion, love, hate and fraternity – to all things they can express. As such, a level of the pure past as all pure differences is expressed in each singular passing present. It then joins all previous expressions as a level of the pure past, one ready to return in another enactment of humiliation and of every other pure difference. An actor, indeed anything in a process of becoming, is a time traveller into the pure past. This is not through a representation of the things in the past, but rather in the way it creates a new level in the pure past at the same time as the pure past engulfs it.

When the present is a dimension of the past the process relating the two is different from when the past is a dimension of the present. With the past as prior, processes of making pass and changing relations in the pure past come to complement the process of contraction in the living present. There is therefore an extraordinary richness and potential for experimentation and applications in Deleuze's philosophy of time. This in part explains the power of his reading of cinema in relation to time, studied in the conclusion to this book. It also explains Deleuze's constant resistance to fixed foundations, objective or subjective, for the philosophy of time and its restriction to a single realm, in mathematics or physics, for instance. Such foundations and reference points cannot account for the manifold processes or for their different sources in related yet different realms such as the pure past, actual representation and the living present. Real time for Deleuze is a manifold of processes fragmenting all subjective or objective grounds into a network of interactions.

According to this philosophy of time, the future also has its own prior processes and includes the past and the present as dimensions. In *Difference and Repetition* this is given as the third synthesis of time. However, like the other two syntheses, the third synthesis is a number of different processes: a cut or caesura, an ordering, an assembly, a seriation and a process of eternal return. The first four are studied in Chapter 4 of this book and eternal return is considered in Chapter 5. Again, I will give simplified versions of the processes here. All of them can be understood from a starting definition of the future as a novel event. This novelty can be seen as necessary within the first two syntheses as that which is distinctive and singular in each living present and in the passing of a present into the pure past. The living present and the pure past depend upon a concept of singularity, or more properly singularities, determining the present and its passing as distinctive through differences that cannot be defined through identities or their negations. Singularities make an event different without having to explain difference through the addition, subtraction or negation of an identity. They therefore avoid the conclusion that there is in fact nothing truly new, only a rearrangement of identifiable components. As in the second synthesis, Deleuze's third synthesis depends upon a notion of pure difference: the new is pure difference determined through singularities.

Once novelty has been defined as pure difference we can begin to see how the notion of radical novelty leads to further processes. A novel event is a cut or caesura in a series because it divides them according to its arrival and its effects on the series. This is relatively easy to understand in terms of large-scale events such as the invention of the atom bomb or the French Revolution, because we can see how such events change ways of living and thinking to such an extent that they form a 'cut' in time. However, Deleuze's idea of the caesura is more difficult and yet also more precise than this. It is not that *some* events lead to such cuts (a thought requiring explanations of why and how we distinguish individual events or classes of events into those that really cut into time and those that do not). It is that *all* events must be such cuts, because to be presents and passing presents they must have singularities or express differences. They must therefore lead to a cut in time according to the effects of those differences, as they pass into the pure past. Everything changes with every new event. The process of the cut is insufficient though, since it leads to the self-contradictory conclusion that everything changes completely with the cut. This could not be the case, because we could not relate before and after the cut and therefore we would not be able to define it as a cut at all. So for Deleuze the future as novelty is not only a caesura. It is also an assembly of events. Each novel event assembles all other events in a novel manner. It travels through them. Every event is novel and it is novel everywhere through this new assembly. This also, though, leads to a process of ordering: we can situate any event in relation to any other as before and after, not through an external reference to their position on a time line, but through reference to the before and after of each cut. The assembly on the other hand implies a seriation, that is, a process running through all events, setting them into series according to the singular work on difference in that novel event.

In order to explain novelty in relation to pure differences, Deleuze borrows from and develops Nietzsche's doctrine of eternal return. Whether this is in fact a positive development is open to criticism and this question will be considered in greater detail in Chapter 6 of this book. Put simply, eternal return is defined by Deleuze in a very formal manner that can be summed up according to the following proposition: only difference returns and never the same. This means that only pure differences return from the pure past to be expressed in novel events. Anything identified as the same, as something that can be the same, can never return. So as a novel event everything travels through time, by cutting it, by ordering it, by assembling it, by setting it into series and by returning through the pure differences it actualises or expresses. Yet no thing and no one travels forward in time as that selfsame thing or person. Instead, travel into the future takes place through the way in which things are made to pass into the pure past. In becoming a dimension of the pure past, in leaving a trace in the pure past as relation of pure differences, present events have a way of travelling into the future as novelty. This novelty is the expression of those pure differences in new events. These too though must perish, as identities and sameness, only to return eternally not as themselves but as pure differences. Travel into the future is therefore manifold. There is travel through the present, when the future is contracted into it. There is travel into the future through the past when the present passes away and can therefore also return as a condition for any novel event. Then there is travel into the future when a novel event selects pure differences to return in a singular way among all the differences of the pure past.

THE CRITICAL POWER OF DELEUZE'S PHILOSOPHY OF TIME

Due to the abstract and rarefied nature of philosophies of time it can seem that they are far removed from practical applications. This remove would be surprising and anomalous within Deleuze's work, given its constant political, social and aesthetic concerns. His philosophy is designed to shape creative and engaged practice. It is necessarily such a practice. We can see this at the heart of his philosophy of time through its commitment to time as multiple processes and to the ideas that times are made with processes and that such processes make beings. They make them become, perish and return not only as active participants in processes but also as passive and cleaved subjects. This gives his philosophy of time a

Introduction

powerful critical edge against objective views, since any situation is much more open and differentiated than it might seem when a unitary and homogeneous definition is given of objective existence. It also, though, gives it a critical edge against subject-based views of time. Time can never originate with a subject (of an action, of a thought, of an intention) because the subject is made and fractured by passive syntheses of time. There is freedom in Deleuze's philosophy of time, not in free will but in a multiplicity of open futures at work in any present and transforming all of the past. The future is not in anyone's hands, since they too are made by passive syntheses of time, but neither is the future closed down and determined. Instead, living with Deleuze's philosophy of time involves a complex and experimental negotiation with many times and with the undoing of strict determination, of subjective grounds and of appeals to eternal values or laws.

For example, the critical force both implicit and explicit in Deleuze's philosophy of time can be shown through its application to contemporary modes of time management and organisation. Take the widespread practice of management through objective setting.[2] This form of time management in relation to organisational power and outputs combines the rhetoric of empowerment through the self-setting of objectives and the fact of heavy-handed control in rewards, veiled threats and real punishments set from the top according to changeable and often obscure goals that hide their broader economic setting in modern capitalism. Recorded objectives, checked over given time frequencies, are designed to bend process to a measurable difference over a fixed time period. More broadly, they are designed to constrict processes in time to a single time line and to a concealment of differences in time through the idea that the time of the setting of the objectives and the time of their measurement are the same in relation to the processes and activities separated by them. The time between the setting and checking of objectives is not meant to exist as something radically open and flexible. Taken to the limit this is even true of the time of the actual reflection and decision on objectives, since there too a complex weave of activity, passivity, process, emotion, bodies, spaces, thinking and desires is reduced to the law of objective measurable outcomes.

Deleuze's construction of a manifold of times, irreducible to one set of laws, or to a single image or space, shows the violence at work in this objective-setting management of time. First, the living present, concentrating past and future upon it, shows the concealment of multiple living intensities in the projection of that present into a future objective. The living present draws the future and the past into it, such that this present is passive to a multiform past, expressed in present tensions and wounds. The present is active too, though not in relation to a single set of objectives, but rather in an interaction with an infinite array of possibilities resonating with that past and those wounds. In Chapter 6 of this book, when we turn to Deleuze's work on time in *Logic of Sense*, we shall see that this means that any activity should rest on an analysis of the wounds and tensions at work in the present alongside a tracing of the work of the past and the future upon them. The setting of objectives is then an explicit attempt to bypass any such analysis through the illusion of a lack of intense wounding, of intense affects, in the present. Deleuze's account of the first synthesis of time demonstrates that past and future collide in many movements determined by singular events. In the teleology of objectives, a supposedly smooth and reliable projection of the past and present into the future makes them the same under a representation of a measured objective and according to general standards. In so doing, it stores up further tensions as objectives are missed, or are perpetually set higher, as they always will be within a model according to debt-based growth, or where they fail either to reflect the richness and fragility of an endeavour or the violent relation between a wider economic regime and the contingent yet truly productive network of living processes, where to live means to become different.

The second synthesis of time is also the basis for a critique of limited and distorting behaviour and theories in relation to time. The passing of the present into all of the pure past as a novel level introduces a responsibility to the past in the present, not as a specific demand from particular past commitments, but rather as an awareness that the present cannot absolve itself selectively of the past. There is no intensity of relations and pure becoming that does not have a call on the passing present. There is no past present that is not connected through the pure past with every other passing present. This does not mean that we should not act selectively and make inclusions and exclusions. It means that none of these is based on a legitimate version of the past, one that could claim that this or that event or emotion should hold sway over all others.

When a way of becoming other is dismissed in the name of a specific historical identity, there is always a distortion of the connectedness of all pure differences or ways of becoming other in the past. It is a denial of its own difference and internal multiplicity. When

Introduction

a claim to exclusive rights is made, or a claim of superior culture, these are always a denial of their connection to what they seek to exclude or override. Does this mean that we cannot struggle against injustices in the name of groups? No. It means that the basis for such claims can never be made on absolute distinctions and final judgements drawn from past evidence or on the absolute nature of present distinctions. They must always be made on the basis of ongoing and relative relations and processes, guided by the communication of all events and all differences, yet also by the necessity of selection across them, rather than by their separation according to essences or pure identities. There could never therefore be a righteous affect (even of anger or love) since it is never expressed alone and it only becomes actual in a present passing away into and transforming the becoming of all affects.

According to Deleuze's work on time, no settled history could lay claim to represent the past. It only contracts the past by inflecting it through singular syntheses in the present. History necessarily changes the past, not through some kind of subjective input that could be countered thanks to pure objectivity, but because the past only exists through processes in the present that make the past a changing event in the present. Does this mean that we are absolved of all claims to facts and objectivity? No. It means that in addition to representing the past in the present event, we must also critically analyse how any such representation is a process of change, a selection and the creation of itself with the past and the future. This analysis will necessarily be a form of experimentation, since there could be no template for the analysis. This is because Deleuze's study of the third synthesis of time and of eternal return teaches us that every event is necessarily novel. It is not new in part, but new through every series of events drawn together in the present as dimension of an open future.

So when the historian claims that the same anger or hatred is at work today as in the past, this is a profound misunderstanding of the affect and of the situations in which it is exercised. Anger is never the same and every situation is different. Does this mean that we cannot compare and that we therefore exist only on minute isolated private islands locked in their times and spaces? No. It means exactly the opposite. We have to test the ways in which our anger is different, understand the ways in which situations have passed, experiment with the ways our own can change. We must hence also pay attention to identity, to sameness and to analogy, not for themselves, but as that which must perish through the work of the new as pure difference. The question 'How is our hatred the same?' therefore misses the deeper problem revealed by a complex and much harder set of questions: 'How is our anger different?', 'Where is a world of identities passing?', 'Who is making it pass?', 'Where should we experiment?', 'How should we create with the new?', 'Who and what is at work making those creations?' Deleuze's philosophy of time is therefore not only critical. It is revolutionary. It is not only revolutionary as a philosophy of time. It is a time of the necessity of revolution, not once, or in one place, but eternally and everywhere.

2
The first synthesis of time

THE LIVING PRESENT

You've been back in the old village a week now. The scars on your knuckles from the old water pump are beginning to heal. The rubber seals on its valves are perished. Early on, each attempt to draw water led to a finger-crushing blow, little usable water and curses of frustration.

Now, your body has learnt to use a short, staccato pumping action. It preserves your hand and yields a reliable if acrid flow of water, as your arm stops short of the rusted metal on each down stroke. The body and brain have absorbed the earlier injuries and later experiments into a trained and automatic action. Failure and pain at the beginning of the week, muscle ache and cautious practice in the middle, have contracted into an unthinking movement, resistant even to your haste in thirst and sickness. Like the sips on water with closed nasal passages warding off gagging on the metallic taste, a smooth and self-enclosed gesture pulls together a series of past processes, some deliberate and others unconscious, such that the current motion is a *passive synthesis* of earlier events.

Habitual gestures such as these support Deleuze's claims about contraction in his account of the first synthesis of time in chapter II of *Difference and Repetition*. Following Hume's account of habit and the role of imagination in drawing together different impressions, he insists on excluding understanding and memory from the contraction of the past in the present: 'It is above all not a memory, nor an operation of the understanding: the contraction is not a reflection' (DRf, 97). In conscious reflection we pull images from memory and analyse them with the understanding. In the experimental moments around the pump, memory of the blows, rapidity of stroke and record of scarred tissue were articulated with the concept of what the safe stroke might be. Memory and understanding worked together as reflection. You essayed the results of the combination of these faculties in slow motion, gradually speeding it up, until satisfied that water could be produced free of bloodletting. For Deleuze, there may well be reflection in the preparation of habitual movements, but it is thanks to the imagination that those preparatory movements are finally contracted together into a movement going beyond each instant of reflection and practice.

How though can Deleuze claim that the past is not synthesised by memory and understanding, since there can indeed be reflection preparing for a trained movement or a novel act? Like engineers designing a water pump, you drew on memories or records of the past, allied them to a current state of understanding and designed a novel movement or apparatus. Does it matter that this can then lead, perhaps only in rare cases, to an automatic and unthinking passivity? This unconscious movement still rests on conscious activity and on records of memory just as much as a new pump design rests on the blueprints of earlier models and on textbooks on the forging of metal, water flow, elasticity of chemical compounds and mechanics of valve action. Why is contraction not found in conscious memories or concepts, but rather in a passive movement guided by imagination?

To answer the question, let us turn to the problem outlined in the opening passages of chapter II of *Difference and Repetition* and to the topic indicated by its title. First, Deleuze has set his account of the syntheses of time within a defence of 'repetition for itself', that is repetition understood not as the repetition of some thing the repetition is 'of', but rather the condition for repetition prior to any consideration of a repeated thing. Under what conditions can we say that there is repetition? The obvious answer seems to be 'when we recognise that a thing has been repeated'. Yet this is the answer Deleuze's opening premise works against. The opening paragraphs of the second chapter raise a paradox for the answer: *there is no repetition until a connection has been drawn between two things*. When two things merely follow one another and no connection is made between them, they remain independent. But repeated instances must be independent, 'by right' Deleuze says, meaning analytically, since each thing could by definition just as well not be repeated. There is nothing in the thing that makes it necessary for it to be repeated or be repeated in a particular way or series. Given any thing, we can conceive of a course of events where the thing ceases to be and is not open to repetition. Many times we have expected a repetition and been disappointed when a part fails or resource runs dry.[1]

So repetition must take place for – or thanks to – something outside the repeated things. This raises another problem. From the outside there must be a difference between the repeated things for repetition to be registered, for without such a difference, there is only one and the same thing and not a repetition. This is a case of Deleuze's reliance on Leibniz's law or the principle of the indiscernibility of identicals: '. . . no two substances are entirely alike and differ only in number' (Leibniz, 1998: 60).[2] Despite the strangeness and novelty of Deleuze's philosophy of time, or perhaps because of it, he is often able to bypass or transform famous paradoxes from the history of philosophy while relying on their premises. *The paradox of repetition is then that although it is defined as the repetition of something that is the same, it can only be the repetition of a difference for something that is not the repeated thing.*

A downstroke of the pump followed by another is not a repetition, until the two are drawn together as in some way different from one another yet repeating nonetheless. Perhaps you notice a slightly increased flow of water, perhaps a little less resistance in the valve, perhaps you are counting the strokes: either way a difference underpins the repetition and that difference is not for one stroke or the other, but rather between the two and for you in the increase in flow or resistance. So if you repeat the movement a third time, this is a repetition of the first two thanks to your expectation of a difference in the coursing of water, or the effort in your arm muscles, or the sequence of numbers, all registered in your mind:

> Repetition (but, exactly, we cannot yet talk of repetition) changes nothing in the object, in the state of things AB. On the other hand, a change takes place in the mind that contemplates: a difference, something new in the mind. When A appears, I now expect the appearance of B.
>
> (DRf, 96)

This raises new questions. In our example, the difference could be seen as primarily in the increased flow, not in the mind, or only in the mind because it is also really in the flow. But Deleuze and Hume's point is that this increase is itself independent of earlier and later moments until it is registered externally. It is insufficient for a conception of repetition until it is connected to the earlier moments in the contemplating mind.[3] Why, though, is a third external term necessary between two things for there to be a repetition? Another question is also raised by our example: could we not include the mind in things repeated, thereby returning repetition to the repetition of repeated things? The answer to this second objection is easier to detect than the first. The inclusion of the mind in things repeated would merely lead to a regress and to the problem of what unites repeated instances in the mind, a problem Hume and Deleuze respond to through reference to the imagination and, as we shall see, Deleuze also responds to through the idea of the living present.

Two remarks allow for an understanding of the relevance of this work on repetition to Deleuze's philosophy of time. First, time is the formal case of the paradox of repetition. Any repetition is also a repetition of the instants identified with it. What draws repeated instants together such that we can speak of time as the synthesis of those instants, if these are in fact logically independent, if there is no necessary internal connection between them, or if, more properly, the very notion of repeated instants implies their independence? *Thus Deleuze is attempting to explain the relation of instants in time, without having to rest on an answer claiming that instants either somehow imply one another or are somehow contained in a larger entity that they are a subset of.* The former possibility would be unsatisfactory for the reason given earlier about repetition: there is no analytical reason why any particular instant should necessarily be connected to any other. The latter fails because we would then have to explain how a property is shared by all instants such that they belong to this wider set, once again setting up a connection between them where there is none to be observed. The second remark is that Hume's reference to the contemplating mind is only a special case of any contraction, rather than the original condition for any repetition. *Repetition does not require a mind, it requires a contraction. Repetition does not take place in time, but rather time – or one of the syntheses of time, the first one – is a contraction*: 'Properly speaking, [contraction] forms a synthesis of time' (DRf, 97).[4] So when we questioned the necessity of mind as a third external term witness to repetition, that was not quite the right problem, or rather Deleuze's treatment moves beyond it by insisting that we do not require a mind as such but a process connecting repeated things. This process is contraction. In the counter-example presented earlier, the response would therefore be that he is demonstrating the necessary contraction between

The first synthesis of time

one flow and another such that there can be a subsequent judgement of 'increase' between the two.

For Deleuze, if we are to have an account of time resistant to the problem of the independence of the instants of time, that is, to the problem of what allows for the connection of those instants, then we must explain how they are brought together in repetition. He calls this contraction an 'originary' synthesis. It is important to distinguish this from an original synthesis. The contraction is not a ground for something else, or a first, or prior, or essential synthesis. Rather it is originary in the sense of giving rise to time; time is made by contraction, it neither pre-exists it nor stands as a condition or container for it. In stating that contraction forms a synthesis, Deleuze is setting time as something formed by a process: a contraction. The distinction drawn between original and originary will be important when we turn to the question of the relations of the three syntheses of time to each other. The first synthesis is originary in that it gives rise to a certain facet of time, but this does not imply that this facet is original or a first foundation for the other syntheses. What, though, guarantees that this originary process cannot be an operation of the understanding and memory? Why in principle could we not give a formal account of all contractions as collections of memories, computed thanks to a given understanding, then leading to a particular action? Why describe contraction as passive and the process of time as a passive synthesis?

A further answer to these questions lies in Deleuze's description of this first synthesis of time as the living present. The contraction of repetitions is a process that gives rise to the living present. Time unfolds thanks to this present, that is, past and future events meet in it, rather than remaining separate entities with no interdependence. In this living present, the past is constituted through a process of retention whereby past events are retained together in the lived present, for instance in the way a stroke at the water pump retains all the earlier attempts it has learnt from. The future is constituted through anticipation. Future events are synthesised by being anticipated, looked forward to or awaited in the living present, for instance, in the way a learning stroke in the present anticipates future strokes and improvements by driving towards them. This leads Deleuze to make the claim that past and future are only dimensions of the living present with no existence distinct from their contraction in it. The future and past as living present become conditions for the past and future conceived as separate from the present, because without the living present they are not synthesised and have no existence as a time that unfolds and coheres. By stretching the present into syntheses of past and future events, Deleuze thus goes beyond the traditional idea that past and future have to be thought from a present instant, the 'now'.[5] The living present passes from past to future, as its dimensions, by synthesising them in retention and anticipation: retention leads into and feeds an anticipation; anticipation rests on and drives off from retention. It is here that we can detect a reason why the synthesis must be passive, since even if some aspects of the processes where the past and future are made can be traced to acts in the living present, these acts themselves depend on passive syntheses far exceeding what could be contained in any one calculation, understanding and set of memories set as conditions for them: '[The synthesis] is not made by the mind, but is made in the contemplating mind, preceding all memories and all reflections' (DRf, 97).

Deleuze's argument for passive synthesis in the living present is therefore an argument about the conditions for synthesis. More precisely it is a deduction of the conditions for particular properties of past and future events in the living present as contraction involving a past and future dimension. An activity, the tensing of a muscle, say, must synthesise earlier movements and later ones. However, Deleuze is interested in a further question about the genesis of that act itself as retention and expectation. The conscious activity and its relation to memory do not contain all the movements, past and future, that it contracts:

> The living present therefore goes from past to future that it constitutes in time, that is as much from the particular to the general, from the particulars that it envelops in contraction, to the general that it develops in the field of its expectation (difference produced in the mind is generality itself, insofar as it forms a living rule of the future).
>
> (DRf, 97)

The particulars referred to in this passage are actual events in the past as contracted through retention. They are particular because they actually occur, as contracted in the living present. Thus, for instance, we could trace back a series of past movements leading to a given gesture. The future events though are not actualised. They are general possibilities (not potentials). Moreover, as passive, they are general possibilities as yet not even conceived of in a mind, but rather set as a general condition for any forward momentum. The important step in the argument is then that any activity, defined as an action with a set of past memories enacted towards a set of future possibilities, cannot include all the particulars and generalities that it retains and expects. Along with Paul Patton's excellent translation, I have kept the French term '*attente*' as expectation here, but there are risks in this from the point of view of a restricted meaning of the term as 'expecting this or that' in the sense of 'conceiving and intending this or that outcome'. It could be better to use 'awaits' rather than 'expects' to give a sense of waiting free of restricted and particular conscious content. As we shall see shortly, this option is supported by Deleuze's use of 'contemplation' in the same paragraph.

Many more particulars and generalities constitute a movement than its definition as an action can account for. Contemplation is therefore not a form of conscious consideration or aiming towards, but rather a form of unconscious receptivity. The active mind makes decisions upon actions, and though this action transfers from a set of past particulars to a set of future generalities, the mind is itself operated on by greater retentions and generalities. As such, *even in activity the present is contemplation, that is, passive absorption and transformation of retained particulars beyond the set considered in an action*. The argument here has many implications. Deleuze's point is that passive contemplation is presupposed by action, since the particular and general selections made by action not only presuppose wider sets they are cut out from, but, more significantly, they are also effects of those sets. What this means is that any action is also a passive retention and expectation, for instance, in terms of conscious action, through unconscious and unconsidered effects, as well as possible yet non-conceived general outcomes.

Conscious activity is only a case of action in Deleuze's argument and definitions. Although his study is framed around the human mind given its context in Hume's work on the imagination, the distinction between activity and passivity does not turn on a distinction drawn between human activity and passivity. On the contrary, passivity and activity are distinguished as processes where passivity is a form of retention and of expectation that is not related through a process selecting particular past events and associating them with a restricted number of general outcomes. This latter process is activity, but passivity is the wider condition for any active process. Passivity cannot itself be an active process, that is, it cannot be determined as a restricted operation from past to future. Conscious calculation is a sub-case of activity. Human passive contemplation is a sub-case of contemplation. A mechanical computation of the best pump design from a group of designs for a particular flow of water given a particular power source is thereby also a form of activity. Perhaps more surprisingly, it also presupposes and is the effect of a form of passivity, whose many types include malfunction under certain conditions, non-computed variations in flows, wear, faulty interaction with other machines and so on. This explains Deleuze's distinction drawn between a passive synthesis and an active one in the mind. A passive synthesis happens in the mind, in the sense that retention and expectation as determined according to paths through particulars and generalities meet in the mind and transfer from past to future. Active synthesis is operated by the mind, made by the mind, in the sense where a property of the mind fully determines the selection of particular parts of the past and possible outcomes. Note, though, that this determination could equally be a property of an algorithm, chemical genetic process or computer program and these count perfectly well as active, rather than passive.

This puts us in a position to think again about the definition and role of the living present in Deleuze's work. We now know that it determines time as a contraction of the past and of the present, but in different ways. Particular past events are contracted into an individual contemplation that is passive rather than active. General future possibilities are prefigured according to this contemplation and an expectation or 'awaiting'. The living present is therefore a process ascribing an arrow to time, from past to future, through the asymmetric nature of the two processes. Why does the arrow move in this direction? Why can't it go from future to past? It is because once the future is defined as the process of expectation, there is no general series of possibilities until we have a process of retention of particulars that then allow for generalisation. The passage is from particular to general and not the reverse: 'Passive synthesis, or contraction, is essentially asymmetrical: it goes from past to future in the present, thus from the particular to the general, and thus orientates the arrow of time' (DRf, 97). Without the prior process of retention we would have no general outcomes for a waiting to tend towards. Given general outcomes do not move towards given retained particulars because then there would be no expectation, notion of the possible, or fan of probabilities, but instead a fixed, though open set of past particulars. Asymmetry, a key term in *Difference and Repetition*, therefore here refers to the essential difference between particulars and generalities, where the former are actual and retained and transformed in the present and the latter are possible and expected in the present. There is no symmetry between the two because any

The first synthesis of time

set of particulars determines a much wider set of generalities, yet also, any given set of generalities neither determines nor includes a set of particulars.[6]

This in turn allows us to discount two interpretations of the living present. It is not a psychological term in Deleuze, as in a psychological state of retention and one of expectation. Instead, both are general and base processes that can take place in many different entities (human, animal, vegetable and mineral, mind, computer, biological system). The first synthesis of time is therefore a very pure definition of the present as process, distinct from the present as present instant for consciousness and from the present as one of three distinct parts of time (past, present and future). Instead, the synthesis draws past and future into the present as two different processes related together in the living present, that is, a process that passes from the retention of the past into the expectation of the future, not as psychological, nor as phenomenological (in the sense of qualities of intention), but as formal processes bearing on different things (particular and general) and setting them into relation. Thus, in the living present, we find Deleuze coining a new usage of the term 'subject', where the subject is no longer the subject of an action, nor therefore the human subject, but rather a passive subject, that is, the subject of a determination that is itself not the active decider or self-sufficient principle for this determination but rather the transformer between past particulars and future generalities explaining this determination through processes exceeding it and that it is passive to: 'Time is subjective, but it is essentially the subjectivity of a passive subject' (DRf, 97). In other words, time is a determination of wider sets and is therefore subjective in relation to an individual determination, but no final explanation or principle of that determination can be found in a particular actor in that process; instead, the central actor – the living present – is itself a passivity and effect of wider processes.

To sum up the argument for the necessity of the living present: it is necessary because in any repetition the repeated terms have no connection until they are synthesised. This synthesis takes place as the first synthesis of time, that is, as the way past particular events reciprocally determine future general ones asymmetrically, or determine each other in fundamentally different ways. Here, 'to determine' means to establish relations through a selection. For instance, when we select a given gesture as the right way to avoid crushing our knuckles on a broken pump we select a path through earlier gestures, which ones are to be repeated, which not, and we select a set of possibilities in terms of expected outcomes, what we expect to happen, what we do not. *Determination is therefore a relating and bringing of order and priority: out of a chaos of unrelated particulars, paths are selected.* These paths then allow for the setting of an order and existence of possible general expected outcomes, while the expected outcomes allow for the setting of the path *but not the existence of the particulars* – hence the asymmetry and the arrow of time. Put very starkly, there is no repetition, no relation between actual events and possible events, and no relation between instants, without the living present. This is a very strong condition in Deleuze's philosophy, to the point where when we spoke of synthesis being external to its terms, this was a question built on a presupposition that Deleuze's philosophy only allows for hypothetically: that repeated things exist outside their synthesis. In fact, empirically we only encounter relations and speculatively we assume that this will always be the case.

SYNTHESIS AND METHOD IN THE FIRST SYNTHESIS OF TIME

There are broader questions raised by Deleuze's introduction of the living present in the opening paragraphs of chapter II of *Difference and Repetition*. These are questions of philosophical method. They can be traced in the closing remarks set out above through the appeal to empirical observation and to an as yet ill-defined term: synthesis. Deleuze's argument is explicitly empirical for the description of the processes of retention and expectation 'When A appears, we expect B with a force corresponding to the qualitative impression of all contracted ABs' (DRf, 97). That processes of synthesis are required is not empirical. It is a logical deduction from the independence of instants. That the processes are conditions for one another in relations of asymmetrical determination is not empirical. It is transcendental, in the sense of the deduction of necessary conditions across different realms (in this case, from actual events to possibilities). However, that there are actual syntheses, rather than just hypothetical ones, is a matter of observation and, here, Deleuze's argument is open to difficult questions. Why depend on Hume's distant observation, rather than contemporary scientific observations (either psychological or in neurology)? Why not turn to the resources of phenomenology, for instance, in Merleau-Ponty's work on perception?[7]

A first clue to an answer can be found in the distinction drawn between empirical and formal in terms of Deleuze's discussion of types of processes. He passes rapidly from empirical remarks to speculative ones about the formal properties of processes and it is this passage that distinguishes his work from more thoroughly scientific empirical observation or phenomenological transcendental work. In the first case, Deleuze is setting out a speculative formal frame on the basis of a sketchy empirical observation. This means that his empiricism combines this observation with the creative construction of a speculative philosophy (with logical and transcendental moves, as we have seen). This partly explains the difficulty of setting down a label for his philosophy: it is empirical, speculative and transcendental. It also invites a deep worry, since there is a danger of failing in each of these moves and standing as poor (unscientific) empiricism, (non-rigorous) phenomenology and (logically deficient) speculative philosophy. There is though a more hopeful counter to this worry. The best philosophy, that is, one that is not stuck with mistaken presuppositions about thought's legitimate status as pure empiricism, phenomenology or speculation, might well be one that uses the resources of all three, on the basis of careful research on them through the history of philosophy, to avoid each one's tendency to impose a view of reality and of thought that is erroneous exactly because it excludes input from the others.

This combination of work on the history of philosophy, empirical observation, speculation, logical analysis and transcendental deduction can be followed in the third paragraph of chapter II of *Difference and Repetition*. There, having established the priority of passive synthesis and the living present, Deleuze works back through his arguments in a reading of Hume in order to show how the separation of instants and of particular past events and general possibilities cannot be conditions for the synthesis. The synthesis is not a synthesis of separate things. On the contrary, synthesis is a condition for the conception of such separation but also for the demonstration of its incompleteness and secondary nature. That is why he starts the paragraph with this difficult statement: 'In considering repetition in the object, we remained short of the conditions that render an idea of repetition possible. But in considering change in the subject, we are already beyond them, in the general form of difference' (DRf, 97). What this means is that repetition cannot be thought of as either the repetition of objects, which explains why Deleuze presented such an approach as leading to a paradox, or as repetition in the subject, which explains why it would be a mistake to associate his reading of Hume with an interpretation of both philosophers as setting down the human mind as the condition for repetition. The Patton translation of these two sentences is therefore somewhat misleading by giving '*en-deçà*' as 'within', rather than 'short of', and by eliding the past tense that makes it clearer that Deleuze is commenting on his own opening paragraph and method.

The thesis that repetition escapes both objective and subjective study shows why Deleuze can appeal neither to brute empiricism, nor to simple phenomenology. This is because both involve presuppositions setting aside repetition and time for themselves. Yet this causes immense methodological problems, since how can we start a philosophical investigation without doing so either on objective or subjective grounds?[8] Deleuze's solution can be found in his combination of methods and, in particular, in its speculative side. As we have seen, Deleuze begins with a reflection on the object but only in order to demonstrate that it leads to a paradoxical dead end when taken purely on its own terms: the object cannot be the ground for a definition of repetition because there is no necessity for repetition in the object alone. For instance, even in an object defined apparently as requiring necessity, in a mass-produced circuit board where the mode of production seems to imply repetition, say, it is possible to envisage that the first real suchlike object off the assembly line could also be the last when the quality inspection notices a fatal imperfection. Deleuze's speculative approach allied to his deduction of transcendental conditions is designed to take such paradoxes as productive for his philosophical thought. As we have seen, he therefore proceeds from an observation of the failure of a grounding of repetition to the speculative explanation of such a failure in a prior synthesis in retention and expectation in the living present. This renders the appeal to the object itself speculative and justifies its cursory nature. There is no need for anything more than a passing study of the object here because what counts is the formal deduction of the paradox, itself inherited from Hume.

The sentence after the statements on objects and subjects testifies to the difficulty of Deleuze's approach, but also to its inherent philosophical values of careful and tentative self-critical enquiry: 'And the ideal constitution of repetition implies a sort of retroactive movement between these two limits' (DRf, 97). Repetition is not objective nor subjective but ideal, where ideal does not means 'of an idea in the human mind' but rather 'of ideal relations as condition for actual differences' (as described in chapter IV of *Difference and Repetition*, 'The ideal synthesis of difference'). At this stage though, Deleuze is only able to indicate indirectly and metaphorically what

The first synthesis of time

this synthesis might be as a 'sort' of movement and 'weave'. It is only four pages later in the French edition of the book that, as we shall see, Deleuze deduces this weave as two-fold relations of difference and of repetition as conditions for integrations and differentiations, as moves to the integral object and to a multiplicity of passive selves. First, though, he proceeds to trace the oscillation between object and subject more precisely, this time not through a study of repetition in relation to the object, but rather in the subject. We have already seen how the subject involved is not the subject of an action. Now, he will show how synthesis as contraction cannot be identified, even after the fact, in memory or understanding in a subject. *Time and synthesis, as well as the living present, cannot be subjective in the sense of properties of the understanding or memory of a thinking subject.*

His demonstration of this focuses on Hume's work on the imagination and draws out a number of key remarks and terms. These are significant because they expand on an earlier puzzle. Deleuze's argument depends on the claim that grounding repetition on the subject involves presuppositions about the form of repetition, but unlike the work on the paradox in the objective approach, we have not seen how exactly. That is what Deleuze will now show. According to his reading of Hume, memory, as represented conceptually, contains particular memories or represented events in their own distinct times and spaces; for example, 'my crushed hand on the pump two days ago' as distinct from 'my healing hand on the pump yesterday'. This memorised past can be distinguished from the past in retention in the living present because in the latter a series of events is synthesised such that they are inseparable. As we saw earlier, the past is concentrated in the living present and does not have a distinct existence. What is more, this concentration in retention must draw past events together because their reality is only through their retention as a series, which is itself a prior condition for any later separation of the series in memory. The same is true for anticipation, where the movement towards a concentrated series of fused general abstract events is separated by the understanding into a set of weighted distinct possibilities. This weighting is done through a scale of probability based on frequency of earlier separate events in memory where, in line with Hume's work on probability as a solution to the problem of induction, something that is recorded many times in memory is given a higher probability.[9]

Deleuze draws two far-reaching conclusions on time and repetition from these remarks. First, repetition implies three moments: a passing away of objective instants due to their unrepeatable nature (a passing leading to the paradox of unrepeatable things); passive synthesis in contraction; and reflexive representation in active memory and understanding. Note that repetition implies all three such that it would be an error to say that it is only passive synthesis: a temptation that must be avoided because it dismembers Deleuze's model and locks it into a focus on a transcendental realm separated from an actual one of passing instants and incomplete represented objects and subjects. Note also, though, that the status of each is different according to an order of priority set according to conditions and determinations (as was also the case earlier in terms of the arrow of time, determined through the relations of particular and general, and past and future). If we insist on one or other of the implied moments at the expense of the others, we miss their relations of reciprocal determination and cut up a philosophy at one of the points where it is insisting on the fateful misrepresentation implied by such distinctions. Second, Hume's study feeds into Bergson's work on memory and on the problem of separate things (each stroke of four bell rings) also being one thing (four o'clock ringing out). The relation between them lies in layers of syntheses and distinctions, all related through conditioning determinations.[10] Orphaned events or the passing instants are the condition for a passive synthesis, which is also the condition for their repetition. This synthesis is itself the condition for a later separation according to representations in memory and understanding, separation which is itself the condition for reproduction and reflection of those syntheses. Each one of these is necessary in Deleuze's speculative presentation which therefore has many methodological facets: *expression* of individuation in the living present (duration in Bergson and imagination in Hume), *representation* of identity in memory and understanding, *creation* of syntheses in a thinking of the relations of the other two and all presupposed ideal relations.

The power of Deleuze's speculative model is set to work straight away and is therefore also tested by him in a comparison of Hume's and Bergson's examples.[11] There are two types of dissimilarity. First, Bergson's example is of a closed repetition (four strikes only) whereas Hume's is open-ended (a series of AB couples without end). Second, Hume's involves cases of AB couples whereas Bergson has repeated undivided elements or strikes. A case involves an internal difference. An element is supposed to be whole. What Deleuze is now able to do, though, is reflect on the significance of these differences on the basis of his model. He concludes that the two examples imply one another. This is because the four strikes also constitute cases, because as four o'clock strikes or unfolds, the first two strikes are an AB couple, then so are the second and third, and then the third and fourth. It is only when the four strikes are over that they can be conceived as a set of four separate elements. The strikes are also open, because the four strikes can be opposed to five, and therefore another AB couple, and so on through all the dimensions of time that include the four strokes. Equally, though, the cases are also elements when they are repeated, when we pass from an AB case to the repetition of two ABs for instance. This latter two-fold AB is itself closed and implied by the open series that exists only in the abstract.

What matters, though, is that Deleuze can explain these relations of openness and closure, and element and case, through his work on time and passive synthesis. The three sides of time – passing instant, synthesised contraction in the living present, and represented instants – are presented in the example of the elements and couples. The element can only become part of the striking of an hour through a passage from instant to contraction, but the synthesised cases imply that each case is a passing instant, and synthesised case and instant can only be represented as closed rather than open series:

> The two forms of repetition always refer to one another in passive synthesis: the form of the case presupposes that of the elements, but the one of the elements necessarily overtakes itself into the one of cases (whence the natural tendency of passive synthesis to experience tick-tick and tick-tock).
>
> (DRf, 98; DRe, 98)

This sentence is instructive for understanding the role of method in Deleuze's philosophy of time. It combines empirical observation ('natural tendency') with Deleuze's transcendental work ('passive synthesis' in the living present as condition for the tendency), with a bold speculative move ('always' and 'necessarily'). His philosophy combines all three methods to provide an explanatory model going beyond the limits of simple empiricism and its difficulties with the Humean problem of induction, while still maintaining an empirical aspect as test and observation. However, it is important to note that the claims of necessity are speculative and open to empirical counters and tests, while the transcendental moves are themselves experimental and grounded in empirical observation. Like his methodology, Deleuze's philosophy is singular and universal, or more precisely, it speculatively oscillates between the two, unable to settle on either one.

Deleuze develops the remarks on elements and cases into a brief study of sensibility and sensation. The empirical natural starting point is in organic sensibility and receptivity, but this is made wider through a series of methodological moves building on the element and case relation. First, he remarks that the relation is different depending on what level it operates on, where level refers to the three sides listed above: passing away of instants, contraction and representation. It will work differently on different levels. When taken at the level of contractions a quality such as the tone of a ringing bell is fused with the contraction of 'elementary excitations' with no subjective input and no conceptualisation. When taken at the level of representation a contracted perception is represented and thereby twinned with an objective quality as 'intentional part', that is as a quality that can be intended by the subject independent of this or that contraction of elements. Thus, on one level, our sensibility is just a contraction of organic syntheses, a series of sensations of warmth, say, prior to and independent of any representation or concept: '[. . .] a primary sensibility that we *are*' (DRf, 99). On another level, though, we are intending beings who can intend towards an object and ascribe a given quality to it, for instance, when we reach out and ask the question, 'Is it warm?' From this point of view, contraction is prior to even that sensation, if sensation is defined as a conscious faculty. This is because we cannot have the sensation in that form until there has been a series of passive organic contractions: 'Every organism is, in its receptive and perceptive elements, but also in its viscera, a sum of contractions, of retentions and expectations' (DRf, 99). The living present and the first synthesis of time determine any conscious representation as necessarily presupposing prior contractions. The definition of the human as rational animal is therefore necessarily non-Deleuzian, not because the human is not rational, but rather because even if the human is rational it must also be a series of non-rational contractions in such a way that reason cannot take the upper hand or fully determine those contractions.

Every existent therefore presupposes passive syntheses. Every existent therefore also presupposes the living presents of each of the syntheses drawn together in it. Since each of these living presents has a past and future dimension, every existent is made of many passive retentions and expectations. Each of these syntheses is a level for it; but also, in a very important term for *Difference and*

The first synthesis of time

Repetition that grows out of Deleuze's work on Proust, *Proust and Signs,* each level is also a sign, that is, a passage from retention to anticipation driving the determination of an existent and its becoming.[12] An organic thing is determined by its syntheses such that they are signs for its future comportment. These signs can be interpreted, but only at the level of representation which must necessarily miss something of the prior syntheses it attempts to read and that constitute it. This forms a productive problem for Deleuze. Given the importance of signs for understanding how we are becoming what we are, and given the impossibility of giving a full representation or interpretation of those signs, what is the right way of living with and living up to the passive syntheses constituting and driving us forward?

We live as time makers – anything exists as a maker of time. This means that the passive syntheses drawn together in any changing thing are processes making time as a living present through that thing. There are therefore many and multiple living presents. There are also many ways of interacting with these living presents and, problematically, whenever we associate them with active representation we capture a side of them and lose another. Following Hume, Deleuze calls this the problem of habit. However, he then notes how habit is often misunderstood due to an illusion coming out of psychology. It is mistake to define habit in terms of our conscious activities, in the sense where we would say, for instance, that I have deliberately acquired the bad habit of using the term 'that is' throughout my text. Instead, for Deleuze, habits are acquired through contemplation, that is, through the passive acquisition of a pattern of syntheses conditioning or determining later activities (where there is a lot at stake in definitions of 'determining', in particular in opposition to 'causing'). My habit of using a particular term to excess can certainly be traced to actions; however, these are not a sufficient explanation of a habit because they fail to explain its relation to unconscious repetitions, retentions and expectations conditioning the habit. This means that learning and unlearning habits must not be seen in terms of the conscious repetition of movement, for instance, but instead must be seen as an interaction with processes that we cannot directly represent or act upon. Here we can see the consistency of Deleuze's philosophy and the role played by his philosophy of time. The first synthesis of time leads to an understanding of the part played by signs in conscious and unconscious habit acquisition. The combination of this oblique form of contraction in relation to signs then guides Deleuze's understanding of learning and teaching, of life as an apprenticeship to signs, for instance as he has absorbed it from his reading of Proust.

OF PEBBLES AND THEIR HABITS

On the east flank of the headland an abrupt and eroded path leads down to a tiny pebble beach. A sickle-shaped indent among sharpened rocks, its smooth stones have been turned to a rare and much prized shape, neither too small to stick to the skin like coarse mud, nor too big to bend soles painfully. Each pebble contemplates the sea and the tides, the currents and the storms, the mass of sister pebbles, flotsam and broken shells. It is a passive synthesis of these events, a contemplating soul ground from repeated washes, like the limpet stuck to its side contemplating it in return: 'What organism is not made of repeated elements and cases, of contemplated and contracted water, nitrogen, carbon, chlorides, sulphates, thereby interweaving all the habits composing it?' (DRf, 102).

Do pebbles really have habits? Do they contemplate the tides? Does a limpet or an oyster have habits? Can they too contemplate grains of sand shaping their shells and, infrequently, the pearls forming in them? These objections to Deleuze's work on the first synthesis of time, and on contraction in the living present, come from at least two opposed directions. First, why speak of habit and contemplation where we have other scientific accounts of the relations between entities, such the concept of cause? Second, if we are to speak of habits and contemplations, should we not reserve these for beings capable of action? The pebble does not synthesise the tides into its rounded shape, but rather the shape is caused by friction, itself caused by tides and currents. The oyster and the pearl are not contemplating the intruders entering the shell. The oyster is caused to react by the foreign body in a way that leads to the pearl. It secretes calcium carbonate and conchiolin protein which over time form a pearl. What need is there for mystical and misleading terms such as habit and contemplation? Why call an effect a habit and thereby hide the cause and effect relation and the many causal laws of nature governing, for example, organic compounds?

Even if we wish to criticise the concept of cause and replace it with laws and probability, with uncertainty and chaotic processes, these too need no unscientific concepts such as habit and contemplation; what would these add to scientific equations and calculations, if not a surplus and inhibiting metaphorical layer, ripe for religious and political mystifications? Would it help a pearl farmer to say that the oyster contemplates the water it bathes in? Farmers require accounts of how the pearl is formed and which environmental states are the most propitious for this growth; they do not need redundant philosophical concepts such as the first synthesis of time and the living present. Or, when sand is shipped in to save a failing tourist destination, should the village mayor read a treatise on the habits of pebbles or a scientific article on the complex science of tides and currents? As a counter to Deleuze's position comes the statement that the habits we acquire unconsciously are better explained as caused or as explained through scientific laws and probabilities rather than by the loose term of habit. For instance, when we learn to walk on the pebbles with the balls of our feet rather than the arches, avoidance of pain gradually causes a change in gait and more supple movements. The painless walk is allowed by a hardening of skin that depends on friction and the layering of dead cells. If something is unconscious, it is not a habit; it is an effect or at the very least an observable pattern. If it is conscious, it might be a habit, but even then only if we think consciousness itself is not an effect.

Deleuze's argument is driven by these objections and he gives voice to them directly: 'This is no barbaric or mystical hypothesis [...]' (DRf, 101). However, he does not explicitly address them critically at this point of *Difference and Repetition* (this comes later, in the third, fourth and fifth chapters of the book). Instead, his concern is to articulate his own position in such a way as to make it immune from these critical questions. The core of his response rests on this statement: 'Habit *draws* something new from repetition: difference (first posited as generality)' (DRf, 101). Deleuze usually turns to italics in *Difference and Repetition* to highlight a key term used in a novel sense (for instance, the term highlighted before this one in the book is 'sign', in its novel meaning in relation to habit and learning indirectly or obliquely). We need, therefore, to decide on the meaning of the term and, perhaps more importantly, on its status. Is it metaphorical? Or is it literal? If literal, is it taken from scientific usage, as terms sometimes are in the book, or is it taken from a philosophical source (such as the use of habit taken from Hume, here)?

A first step in deciding on an interpretation of Deleuze's use of the verb '*soutirer*' (to draw) is that it is not metaphorical. The verb does not stand for another process it is meant to allude to or represent, but rather habit is a process drawing on repetition.

What, though, does 'to draw from', or 'to draw out' mean here? In a preliminary sense, it is to draw out difference from a repetition. We know this not only from the statement, but also because habit has been defined in relation to repetition which we know from the preceding paragraphs must involve a difference. We also know that this difference lies in a synthesis or contraction of a series. So habit is about the creation of a difference but where the difference itself cannot be a represented identity. Instead, following Deleuze's work in the previous chapter of *Difference and Repetition*, 'Difference in itself', this difference must be a varying relation, rather than a fixed quantity or quality, or an identified and limited body. Habit draws a differential variation from a repetition. It does not do the same thing, as the commonsense understanding might lead us to assume, but on the contrary creates a change or becoming in the series.

'*Soutirer*' is a technical term with a chemical basis from winemaking. It is one of many taken from chemistry and biology used by Deleuze when he wants to point to this differential variation (at other times he uses examples such as ebullition, at DRf, 296, for example). The term means to draw wine from one barrel into another, for instance, in order to remove sediment. It is important, however, not to identify the concept with the casks, or the wines in their apparently fixed states in each one. The process Deleuze wants to map his philosophical concept on is not the passage from one state to another. Instead, Deleuze is interested in the process itself and, more precisely, in the introduction of a difference in intensity, a differential variation, in the process synthesised as time. So habit is a contraction, not in the sense of a passage from a dilated to a contracted state, as Deleuze says about heartbeats, but rather a synthesis of events (contraction and dilation) as a differential, an ongoing variation of intensity or a becoming – and not a difference between two states. So we can now better understand what habit is as retention and expectation: it is the synthesis of a variation in intensity over events, where retention is the absorption of past variations and expectation the impulse to future ones.

However, is not this appeal to a term from winemaking and other processes of drawing metaphorical in exactly the way denied earlier? It is here that we need to return to Deleuze's method. He is not using the verb '*soutirer*' to represent something else, but rather taking an observation of the process of drawing and constructing a novel philosophical concept from it. 'Drawing from' means the synthesis of a series in a novel manner, such that differences in intensity appear within the series and contract it differently in relation to other series. The new barrel changes the relations in intensity to earlier ones and later ones. It changes the relations to our noses and palates, the relations to the crushing of the grapes, the soil where the vines grew, sunshine and rain, pruning and training. This explains the importance of the living present as synthesis: the novel reaction is the living present as a contraction of all the series around it into something new where they are retained differently and lead to different expectations or forward momentum.

Deleuze's answer to the critique based on cause and effect is therefore that the process he is defining and describing is not about associating identified causes and effects repeating in the same way over time. Instead it is about a novel variation continuing to vary, thereby constituting time as the synthesis of the variation. The synthesis covers or includes elements we would usually associate with the cause and the effect, so instead of a cause associated with an effect, we have a novel synthesis that changes all the elements and cases of a series. If we draw a wine from one barrel to another, we can identify a causal relation between tannins and astringency in the wine and explain that wine will always be less astringent or tannic if an amount of sediment is left in the first barrel. Deleuze, though, wants to explain something different and that is the way in which a variation in intensity changes past and future relations through all series stretching out from the living present. For instance, when a wine creates a singular delight or disgust, this novel intensity carries through all the series coming together in the present singularity expressed on the palate of the taster. The contrast can be thought of as the distinction between an explanation of why things remain the same over time and an explanation of why they vary. A causal explanation, for instance in terms of a law applying to particular instances, accounts for a high probability for a specific outcome (or certainty in some versions of causal explanations). An explanation in terms of the first synthesis of time accounts for a novel state of a series through all its elements, not in terms of an invariant such as a law, but rather in terms of a difference, a novel intensity or variation. The contrast can therefore also be thought of as the difference between the conditions for similarity within a structure and variations in a system.[13]

This allows us to consider another important question about the relation of the two explanatory structures. Should we think of them as 'either, or' options, where we either have an account consistent with philosophical naturalism and hence one that follows the latest science, or one based on Deleuze's work on the conditions for difference in repetition? The answer to this question is that Deleuze's work combines both positions. It does so for the important reason that without sameness, for instance as captured in reliable relations of cause and effect, Deleuze would have no actual events to refer to and he would fall into the trap of a world of pure becoming and the paradox that if all is becoming then there is nothing to ensure continuity of reference through time. In *Difference and Repetition* Deleuze considers this paradox through a study of Plato's *Theaetetus*, from a philosophical point of view, and through an analysis of the concept of disparity, from the point of view of the sciences (in chapter V of *Difference and Repetition*). From both angles, he does not seek to deny scientific evidence and theories, but instead seeks to complement them with an account of the role of difference as taking a primary but never complete role in relations of determination between actual identities and ideal differentiations. We can and should consider an event as the referent of scientific accounts. However, these accounts are incomplete unless taken with a more speculative model explaining the intensive difference making each event different.

Deleuze's use of the concept of soul is consistent with Hume's use of the term in relation to the imagination and the effects of distance and contiguity: 'Since the imagination, therefore, in running from low to high, finds an opposition in its internal qualities and principles, and since the soul, when elevated with joy and courage, in a manner seeks opposition [. . .]' (Hume, 2009: 278). This adoption of an outmoded term must not be interpreted as a return to a theological or philosophically obscure set of ideas. Instead, 'soul' has a precise meaning in his metaphysics. The soul is the intensive difference contracted by a habit. It is the difference allowing a series of events to be synthesised in a living present, as different from identifications and representations of sameness to other events. The soul of any thing is therefore the singular way in which it contracts past and future series. This does not mean that the thing does not also have an identity that can be referred to and treated according to causality, reliable scientific laws and probabilities. It is rather that the soul explains why a thing is not only such an identity. It is also why Deleuze insists on the singularity of each thing, where no two grains of wheat or of sand are the same. Thus the soul is to be associated with a process of individuation and any thing has a soul, because as we have seen in Deleuze's study of repetition, any thing must be a repetition of difference (of difference in itself or intensive difference).[14] Yet, since this synthesis is passive, we must

The first synthesis of time

also say that the soul is contemplation rather than action: 'We must attribute a soul to the heart, to the muscles, the nerves, the cells, but it is a contemplative soul whose role is to contract habit' (DRf, 101).

This last claim with respect to the soul is interesting for two reasons. First, it stresses the multiplicity and lack of hierarchy of Deleuze's philosophy. Deleuze's world is radically multiple: it is constructed from multiple and irreducibly different syntheses forming many different perspectives on one another (where perspective is a way of describing different syntheses and contractions). A beach is not a totality of beings. It is a multiplicity of contractions which cannot be organised into a final order, logic or pattern without imposing an illusory sense of the real. However, the second reason the earlier claim is interesting demonstrates a more difficult aspect of this multiplicity. It leads to a problem of selection with respect to perspective: which one should we select? Which soul matters in relation to a given task, that is, which one are we going to take as primary or the one where we begin an enquiry? Where are we to assign value in a given series of events, with the human heart, the stones on a beach, singular and incomparable grains of wheat ('There is a contraction of the earth and of humidity called wheat, and that contraction is a contemplation, and the auto-satisfaction of that contemplation' (DRf, 102))?

The multiplicity of contractions and contemplations is the basis for Deleuze's redefinition of the self away from the self as subject of actions and towards a multiplicity of passive selves underlying and constituting the active self. Instead of thinking of the self as a single self for each individual, traceable back from the subject of actions and open to reflection and self-representation, we have many selves that are all different syntheses of one another. We are a multiplicity of passive contemplations as conditions for the active self we subsequently ascribe to ourselves: 'Under the acting self there are little selves which contemplate and render possible action and the active subject. We only say "me" through those thousand witnesses contemplating in us; it is always a third party who says "me"' (DRf, 103). However, in the same way as identity and repetition of the same still play roles in the complete process, action is still necessary and primary in integrating those multiple selves. Without reference to an identified acting self we would not have a principle for associating the prior multiplicity. There are therefore two types of contractions: there is contraction in relation to generality where the act takes particulars and integrates them into generalities, for instance, when we decide to pursue a given course of action against a background of multiple desires and pressures, or when a severe storm alters the capacity of a beach to support particular types of organic life forms; and there is a contraction in relation to contemplation and passivity where identities are differentiated into a multiplicity of syntheses or selves. These processes of integration and differentiation, of passivity and activity, are necessary in relation to one another. Integration brings individuation to a multiple chaos, but differentiation is the condition for individuation in two ways. It is what is integrated and the condition for any integration, in the sense where any integration itself depends on prior differentiations.

This puts us in a position to interpret another stressed statement from *Difference and Repetition*: '*Difference is between two repetitions*' (DRf, 104). The Patton translation renders this as 'Difference lies between two repetitions' (DRe, 76). But there is a risk in introducing a verb of position ('to lie') since it prejudges the question of whether difference is situated between two repetitions, or whether it is the repetitions that create difference between them. This is important since in the first instance difference can be thought of as a substance, or a zone, or a realm, an entity of some sort (even if multiple) independent of repetition. However, in the second, difference is in the relations of the processes of repetition themselves.

This second option is the better one not only on textual grounds, since Deleuze has described difference as 'inhabiting' repetition, but also on philosophical grounds, since it allows us to explain how the two repetitions – integration and differentiation – relate to one another. They both introduce difference into the other. Integration goes from a chaos of passing instants to activity and representation, itself dependent on passive syntheses. So it goes from the paradoxical multiplicity of instants that was presented earlier in the chapter to two differences: difference as opposition between represented identities, but also and primarily, difference as passive synthesis. This second difference is differentiation, where any integral thing is undone into a multiplicity of passive selves or syntheses in time. The living present is both repetitions and both these repetitions are also '*between two differences*' (DRf, 104) since the repetitions are nothing but the relations of two processes of difference, or where difference is created. The first synthesis of time is therefore a differentiation and an integration, a contraction allowing for action, and a passive synthesis undoing that contraction and opening up to novel differences: 'And already originally, the generality formed by the contraction of the "tick" is redistributed in particularities in the more complex repetition of the "tick tocks"' (DRf, 104). The primary repetition in the first synthesis of time is hence in passive synthesis, in the renewal afforded by a differentiating synthesis that means that no process of integration is final, determining of a complete entity, or free of internal differences and differential intensities.

THE PASSING PRESENT

The injured pigeon flutters for a while in ill-shaped curves; then it falls to the ground. Even there its beating heart and reflex movements drum on the earth. Once all living signs cease, other syntheses come to the fore, parasites and microbes dismember the carcass. A rat drags some of the flesh away. Leaves cover the remaining traces and the bird becomes only a minute variation in the mulch, until the living present, the multiple syntheses integrated through the pigeon's actions and the multiple disjunctive lines made through its passivity (the parasites, the rat, the land fertilised by its droppings, the small variations in currents interacting with its wings) all pass away. The integrated bird is gone and with it, not only all the events working through its passions, but all those events in their interaction with its acts, its actual integrity.

After his explanation of the interaction of differences and repetitions in time, in the first synthesis of time, Deleuze then proceeds to another bold speculative move. He has shown how the past and the future are dimensions of the present, because they are concentrated in its syntheses. He has also shown therefore how 'only the present exists' because it does not itself take place in another time, but rather time is made in the living present (DRf, 105). However, the existence takes a special form presenting great difficulties for the argument as it stands at the point in *Difference and Repetition*. Here is Deleuze's statement of this difficulty: 'Nonetheless, this synthesis is intra-temporal, which means that the present passes' (DRf, 105). The Latin prefix is crucial here. It cannot mean that the present passes in another time, since this would be a direct contradiction of one of the opening premises of Deleuze's paragraph (only the present exists). But if it does not mean in another, it must mean – more correctly – within time and hence within itself.

Yet if it means within itself, it can either mean that the present passes into the present, a dull contradiction, or it can mean that the present passes into one of its dimensions, an interesting but technically very challenging idea. It is challenging because it raises awkward paradoxes. If the present passes into the past that it synthesises, there seems to be a problem of succession: the present passes into that which it has first synthesised and therefore transformed. The synthesised past is no longer there to pass away into. If the present passes into the future that it awaits, there seems to be a similar problem of reversed succession: the present passes into a future that cannot be yet, since it is anticipated or awaited. The paradox turns on a previous aspect of Deleuze's philosophy of time that we studied in earlier sections. Time has an arrow passing from past to future. If the present passes into one of the dimensions produced from its syntheses as living present, it either goes against the arrow back down the synthesis, or it goes before the arrow, prior to the synthesis of anticipation. The metaphors of back down and go before are not intrinsic to the paradox here. More formally it can be described as a problem generated by the asymmetry of time. If the present passes into its past dimension, it must change the particulars it has already synthesised.[15] If the present passes into its future dimension, it carries actual particulars into the generality of the future, thereby contradicting its definition as only generality. In the first case the arrow of time must be reversed. In the second, it is denied.

Deleuze's arguments for the necessity of the passing present, against these paradoxes, marshal nearly all of the prior concepts and principles of the first synthesis of time. He sets them against a counter-position, but also adds new concepts and principles. The counter-position that appears to bypass the paradoxes of the passing present is to consider the present as perpetual: 'We can without doubt conceive of a perpetual present, a present coextensive to time; it is sufficient to apply contemplation to the infinite of the succession of instants' (DRf, 105). A present would not pass, if it synthesised all instants in one go, if the synthesis was therefore not a stretch or duration but rather an instantaneous contemplation, once again reminiscent of Augustine's arguments on human and divine time in the confessions, where divine time does not pass. The present that does not pass would be the present of a god, contemplating all in one.[16] However, Deleuze's argument is different. It turns again on empirical beginnings, since he claims that infinite contemplation is not a physical possibility. This basis is very thin at this point, since it moves straight away into a series of deductions and speculations about necessity that go back to the discussion of elements and cases in repetition (following Hume and Bergson, as discussed earlier on in this chapter). Synthesised elements must necessarily pass and each case that is synthesised is already a passing between its components. Four o'clock strikes because each blow has passed. Tick-tock implies the passing of tick and the arrival of tock. Physical duration is therefore the stretch that it takes for a contraction to pass. Durations are therefore multiple and overlapping: 'An organism has a present duration, diverse present durations, following the natural scope of contraction of its contemplating souls' (DRf, 105).[17]

The multiplicity of syntheses in the living present is therefore matched by a multiplicity of durations, themselves defined as passing away. This allows Deleuze to introduce new terms that we would usually associate with either physics, or psychology or phenomenology. He, though, combines quite minimal observation with careful deductions and bold speculative moves. Exhaustion and fatigue are necessary aspects of the first synthesis of time as passing present. Both are deep concerns in Deleuze's work and mark his readings of Beckett and his broad understanding of life. They are not, though, simply empirical physical properties such as the metal fatigue occurring probabilistically over a series of cycles of loading of a weight-supporting beam, for instance. They also are not simply psychological states, whether shown behaviourally, through introspection or through more objective observations of muscle or neural activity. Finally, they are not phenomenological states associated with the conditions for intentionality, such as boredom. Instead, fatigue follows from the passing present. It is the fading of contemplation and synthesis as earlier events in a contracted series pass away. As a principle essential to time, fatigue cannot be denied and resistance to it must take account of its necessity and temporal form.

Once again, is Deleuze giving us a dangerously superfluous speculative and metaphysical account here? When an engineer seeks to calculate the gauge of a rail, given metal fatigue, is reference to the passing present necessary? Or when we try to find ways of living with mental or physical exhaustion, should we pay attention to the first synthesis of time? When we try to design sustainable farming techniques, are the contractions implied by the living presents and passing presents of the soil required reference points? The answer is yes, but only in careful interaction with the natural sciences and with the arts and humanities. Deleuze's work on time provides a critical and creative set of principles to add to and contrast with other approaches. He shows this critical side in a discussion of the relation between need and fatigue, immediately after his discussion of the lasting present:

> That's why a phenomenon such as need can be understood as 'lack', from the point of view of the action and active syntheses it determines, but from the point of view of passive syntheses conditioning it, it must on the contrary be understood as an extreme 'satiety' or 'fatigue'.
> (DRf, 105)

When thinking about need we must view it in terms of the integrations it involves, how actions draw together and synthesise wider series and therefore lead to a general expectation. The synthesis and the integrated subject of actions in a process of becoming determine a need that can be mapped on the passage from a series of contracted events and a general expectation (*I need coffee in the morning*). But this is not a complete picture. The much broader, and in principle unlimited, passive syntheses determining the action and the thing as becoming something different according to multiple contractions cannot be understood as determining lacks, but rather satieties, that is, syntheses passing away and becoming redundant (*Too tired even for coffee*).

So there are precise principles that can be taken from Deleuze's work on the living present and the passing present. It is not enough to think of action in relation to a set of needs, the requirements of a material, of a body, of a mind. Each one of these is not only needful but also necessarily tiring, not in the sense of requiring the same replenishment, but rather an awareness of how some of its needs have become satiated and fatigued in the sense of lost and past. On a human scale, Deleuze develops this in relation to signs and fatigue.[18] We have to act in relation to the multiple active and passive syntheses, the contemplations, associated with our living present: 'All our rhythms, reserves, reaction times, the thousands of weaves, of presents and fatigues we are composed of, are defined from our contemplations' (DRf, 106). Any act seeking to move without taking account of these, or by 'moving faster than them' is making a mistake. This is also true of signs, which lead to a wisdom Deleuze ascribes to the Stoics, such that a wound is not the sign of a past wound, but of a present series of passive syntheses.

We misunderstand a sign when we think of it as referring to a past event or to a future one.[19] A sign is always a present event and has to be read in its present syntheses, contemplations and concentrations (*How is my coffee habit retention of past events and waiting on future ones?*) So, again in response to objections around the redundancy of Deleuze's philosophy of time, we find him making an important practical distinction with respect to signs and how to read and act upon them. We must distinguish natural signs from artificial ones.

The first synthesis of time

A natural sign relates to a present, to the work of present passive syntheses in a living present and in relation to the passing present. An artificial sign refers to the past and to the future as distinct from the present (*Imagine your life free of your coffee addiction*), for instance when we ask ourselves abstractly what we want to be, before we seek out the signs of what we are becoming in the present, before attempting to learn our present signs. Artificial signs stop us from learning and turn us away from the natural signs that can help us live with our wounds, fatigues and multiple becoming, because they draw us to a past severed from its work on present wounds and to a future cut away from its pull on present expectations.

Deleuze develops these practical remarks further in two areas: the relation between need, habit and questioning, and the passage from passive selves to larval subjects, emerging singularities determining a novel becoming, in relation to fatigue and passion.[20] When need is conceived as lack, and hence in a structure of negativity (*Coffee is what I need and do not have*), it is misunderstood as a relation to passive syntheses which are transforming the need through its passing away and many forms of fatigue (*Coffee just does not have the same effect any more*). There is no questioning in a structure of negativity, because we presume to know what we need and hence the sole difficulty is how to get it. In the complete sense of our presents, as becoming and falling away, need is more than this negativity, because it is a sign of the wearing away of lacks and the appearance of novel expectations. Yet, since this passing away and moving forward in the living present are passive syntheses, we cannot directly represent or know them. They are therefore sources of questions, defined in terms of deep-seated problems rather than simple and readily available answers:

> Is it not proper to the question to 'draw on' an answer? The question presents at once the stubbornness and obstinacy, and that lassitude, that fatigue, which corresponds to need. What difference is there . . .? Thus is the question the contemplating soul asks of repetition, and that it draws from repetition.
>
> (DRf, 106)

Here, Deleuze returns to two ideas developed a few pages earlier and studied here. A question draws on ('*soutire*') an answer, that is, it refines and selects within it, sets it within a series of repetitions. More importantly, it does so by experimentally searching for a novel difference that has passively appeared within the series; this novel difference is a larval subject driving towards a further synthesis of the series. It is not an active subject, or one with a set identity; it is rather the emergence of novel singularities against a background of multiple passive syntheses and series.

There is therefore no pre-set answer to a question when it is defined in relation to a problem and to need. On the contrary, the commonsense answer, set in categories by a good sense, must be transformed and drawn on in order to allow difference to emerge. Against the obvious answer, and exactly because it is not an agreed and categorised answer, the question searches for a way to follow on in the wake of multiple passive syntheses, living with fatigue and expectation, transforming those series again. That is why the question must be situated in relation to habit, signs and learning. It works within habits, responds to signs and experiments with ways of learning with them, an apprenticeship to one's own signs. All of these rest on Deleuze's philosophy of the first synthesis of time: 'A first question-problem complex, as it appears in the living present (the urgency of life), corresponds to the first synthesis of time. This living present rests on habit, and with it so does all of organic and psychic life' (DRf, 107). Habit as defined on the basis of Deleuze's work on the first synthesis of time is itself the process where the syntheses of the passive self, 'the world of the passive syntheses constituting the system of the self', are also larval subjects, that is, the multiple subjects of actions prior to reflection, representation and understanding. Deleuze's philosophy of time allows him to turn philosophy away from the opposition of passivity and activity, to an understanding of life – of all things that become – as activity drawn from passivity.

3

The second synthesis of time

A TIME WITHIN WHICH TIME PASSES

In *Difference and Repetition,* after his study of the first synthesis of time, Deleuze moves on to the second synthesis which we can loosely describe as a synthesis of the past. This synthesis stands as a condition for memory. Like the first, the second synthesis is passive and it is also a condition for active syntheses. Time is therefore manifold for Deleuze, with different syntheses interacting in a fractured and complex manner, allowing for dislocations and changes in perspective. These fractures and rifts are constitutive of his philosophy of time. Time is radically fragmentary; as are those processes making and made by time. Much as things exist as time-makers, they also exist as made by fractured and dislocated times, themselves made by other processes. Time and process coexist like reflections in a hall of broken mirrors, offering multiple perspectives to follow and recreate, but never a full image.

When he passes from first to second passive synthesis, Deleuze straight away tackles the difficult question of why a second synthesis of time is necessary. This is important, because it could seem that given the past and future as dimensions of the present, itself defined as a contraction of time, there is no need for any further times. It also invites the question of the relation between the two times. Are they dependent upon one another? If so, how do they interact? Does one take precedence over the other?[1] Is there an order of times, perhaps a priority of the present over the past? More precisely and more awkwardly, what is the relation between the second passive synthesis of time, defined as the synthesis of memory, and the past as dimension of the synthesis of the present? What is the difference between active and passive memory in this contraction of time? Finally, should the two times be considered distinct, and if so, exactly how distinct, or should they be seen as part of one and the same broader time that includes both of them?

The first step in Deleuze's argument for the second synthesis of time is based on his work on the first and returns to three features we noted in the previous chapter: *the first synthesis is originary but not a pure origin; the present in the first synthesis of time is a passing present; the present is constituted of many durations or stretches that overlap.* The first of these features sets the scene in terms of the relations between times and introduces an important argument on foundations to follow later. The first synthesis is a process that makes time, it is originary, but that does not mean it is original, in the sense of a pure origin, and independent of any other time. This therefore leaves it free to be founded on another time. It also means that the question of foundations and structure of these times will have to be considered in depth and explained further, in order to explain how an originary time depends on another. The question of foundations is critical for understanding Deleuze's work on time, but more importantly for applications of that work. Despite its status as an originary process, the present in the first synthesis of time is not eternal, it passes and it passes in a very particular way. The present of the first synthesis of time is a multiplicity of syntheses, of stretches or durations. This leaves open the difficulty of how all its stretches are related, a problem concerning the wholeness and completion of the synthesis, but also concerning order of priority and interactions in time. We saw this latter problem in the problem of selection in the previous chapter: is it contingent as to which duration or synthesis we begin with and relate to all the others?

Once again, the key statement in Deleuze's argument is italicised in his text: 'We must not reject the necessary conclusion: *there must be another time in which the first synthesis of time operates*' (DRf, 108). The statement is based on a paradox, itself drawn from the three features outlined above. As we saw in the deductions in the first synthesis of time, paradoxes have a two-fold function in Deleuze's work. On the one hand, they have a critical function demonstrating the limits of given claims and positions, most often those based on common sense and on hidden presuppositions about identities. On the other hand, paradoxes have a generative function, that is, the paradox generates a problem which itself leads to series of crea-

The second synthesis of time

tive and speculative partial solutions, where partial means that the problem recurs but transformed.

The paradox of the present is 'to constitute time, but to pass into this constituted time' (DRf, 108). It is paradoxical because it sets up a fork of two contradictions. How can we pass into something we have constituted, if we have to constitute it before we can pass into it, yet if we are already passing away while we constitute it? If the present passes away first, then it cannot constitute the past, since the present is already gone and a new present must constitute the past. If the present constitutes the past first, then it must pass away into a past that it has not constituted, since there will be an interval, a difference, between the constituted past and the one the present passes away into. *Drop a leaf into a river (constitute the past), then drop another leaf (pass away) and the two leaves will remain separate, thereby contradicting the idea that you constituted the past as you passed away.* It could be answered that passing away and constituting take place at the same time, but this would not be satisfactory because Deleuze's account of the first synthesis of time sets it out as a contraction involving dimensions that are not simultaneous, a stretch or synthesis of the past and a synthesis of the future. Perhaps then it could be answered that though the syntheses are not simultaneous, they are indivisible. Therefore constituting and passing away cannot be treated separately (*there are never two distinct thrown leaves*). This answer is unsatisfactory because it would imply that no present ever passes because it is eternally prolonged into the future it expects, which would contradict Deleuze's arguments against the eternity of the present in his work on the first synthesis of time and his account of the element and case structure of synthesis, as shown in the previous chapter.

Deleuze's answer to the paradox is to separate the time into which the present passes away and the originary first synthesis. In a characteristically rapid move, he therefore claims that there must be another time in which the first operates. There is a translation issue here since in the current translation this is given as 'another time in which the first synthesis can occur' (DRe, 79). The use of occurrence is not quite right, since it loses the process-like quality of the syntheses of time. It is not that they happen, but rather that they make and do. This is important because it could seem that the first synthesis simply passes away into another time as locus or unchanged receptacle. That is not correct since he is going to show how there is a transformation in this passing away of the first synthesis of time with the second. It is a two-way reciprocal determination, rather than a passing into an inert and unchanged medium or collection. Both syntheses of time are active in many different ways and passivity must itself be understood as a process in Deleuze, rather than inertia or indifference. So the meaning of 'to pass away' should not be seen as an inert falling into disuse. It is quite the contrary. *To pass away is to pass away in a synthesis of the past as memory defined as the second synthesis of time.* Reciprocally, though, this second synthesis will have a return determination on the first. We shall see this in the next sections when we discuss memory or pure past as completed by the living present.

It is now possible to observe two important principles of Deleuze's speculative metaphysics: he does not allow for nothingness or a void as an explanatory principle; all processes are at least two-way, but as such they are asymmetrical and, even where they appear to be symmetrical as in the case of causality, it is because they are considered incompletely. We can see both these assumptions at work in the passage from first synthesis of time to the second. In deducing this passage, Deleuze refers to necessity in two consecutive sentences ('necessary consequence' and 'necessary referral', though the second is omitted from the English translation). However, this is no simple logical necessity, since it is does not follow from formal logical operations. Rather, it is speculative because it follows logically only if we accept the structure of a complex speculative metaphysics. This shows strongly when we voice the objection to the deduction that the passing present does not necessarily have to pass into anything. Why could it not simply fall into nothingness or be voided? Such an option is never considered as valid within Deleuze's metaphysics. When things pass, they pass into something else according to a process: *the first synthesis of time must operate in another.*[2]

There are at least two ways in which this premise can therefore fail. First, if there is empirical evidence of the kind of void or nothingness that would make a reference to passing away into something unnecessary – memory, in this case. Second, if another metaphysics including nothingness provides an explanatory structure that is more creative (intense and enriching) and more satisfactory (more consistent and more comprehensive). For the first, passage into complete nothingness seems unpromising, though perhaps some cases can be found in cosmology – black holes, for instance. However, it might not be necessary to seek out a case of absolute nothingness or pure void, since given Deleuze's characterisation of the second synthesis of time as memory and given his requirement

The second synthesis of time

for two-way processes, evidence of annihilation of things seems to offer a good counter to his position.[3] Why say that it is necessary to assume that there is a second synthesis of time as memory into which the first must pass, when contractions such as a beating heart stop, rot, disintegrate and disappear not into memory but into ashes and far-spread organic matter? This means that in following Deleuze's work on the second synthesis of time, it will be important to find answers to the objection that present syntheses do not pass into a second synthesis but are annihilated, at least as far as any two-way process or active return of them is concerned. The passing of time would then not be a passing into memory but a passing into oblivion, joining the forlorn ranks of billions of dead and forgotten beings.[4]

The next steps in Deleuze's argument appear doubtful from the point of view not only of this objection, but also of the second concern listed above on whether Deleuze offers us the most robust metaphysics. It is at first sight dubious because he uses what seems like highly rhetorical and metaphorical language to draw a distinction between two processes: foundation and founding.[5] Foundation is described as 'concerning the soil' while founding 'comes rather from the sky' (DRf, 108).[6] Nonetheless, there is a rigorous interpretation of the distinction; it begins with the earlier originary and original division. The first synthesis of time is a foundation because it is a process of 'occupation' and 'possession', that is, it determines an open space according to patterns and to differences. For instance, the animal marking of a territory or the occupation of a patch of earth by weeds is a process of foundation. A space is occupied by a repetitive pattern. It operates through repetitions and habits, and therefore works through the first synthesis of time as living and passing present. Yet the foundation is still confused. Since the durations of living presents overlap, and since they all pass away, the form and determination of that passing away are still not given. The animal pushed from its territory by a predator and the plant overtaken by a hardier weed seem to fall away together into an indeterminate chaos, a nothingness exactly in opposition to Deleuze's concern with memory. Yet, each can also be recalled, so what is the condition for that recall of past things, for the fact of such recall?

Deleuze's preliminary answer is that this must be a process that determines the 'property' and 'appropriateness' of foundation in relation to a summit (DRf, 108). It shows how a passing present belongs with others. What this means is that founding determines the relation between the passing presents and living presents. Instead of having a perpetually shifting ground, or one that remains eternally the same, the second synthesis assigns proper well-determined relations between passing presents. It does not make them pass by engulfing or consuming them, but rather makes them pass by giving them a well-determined relation to what is to come. This can be understood through an example of two opposed procedures. Two monasteries collect books from two rival cities whose conflict threatens their records of their past victories and of their cultural achievements. One monastery, in the grips of an extreme piety, views all human records as a sinful attempt to usurp the divine function of final reckoning. It therefore incinerates every book bequeathed to it, ditching hot ashes down the cliff edge it stands upon. The other order sees human memories as a way to honour the divine plan. It therefore constructs a great library, carefully organising books according to date and subject, in a vast edifice destined to outgrow all other monastic buildings. After the battles, one city will have seen its history pass, but only in the sense of destruction and forgetting. The other has also lost earlier times into the past, but this time they have been made to pass according to a determination of a proper relation of past to present. For Deleuze, only the more tempered holy order has truly made the present pass through a founding, because each present can only pass if it already has a determined relation to other presents, and this is provided by its passing away as given as appropriate rather than consumption into oblivion.

There is an important distinction to be made here in reading Deleuze's concept of foundation, defined as making things pass in a determined and appropriate manner. It should not be understood as meaning that there is a single proper and determinate manner, free of the test of doubt and standing alone as the one true foundation. This is a view of foundation in philosophy running counter to his speculative and inherently multiple approaches. *The necessity of foundation is only a requirement for a determination of appropriateness, not the final complete determination.* This is a far-reaching point because it sets up oppositions to other philosophical positions, for instance, to Cartesian foundationalism;[7] yet it also sets very difficult restrictions on Deleuze's work on the second synthesis of time. This is because the second synthesis of time as that which makes the present pass through a process of foundation cannot depend upon or lead to a finally established and identified true foundation. It has to allow the present to be determined differently as proper and appropri-

The second synthesis of time

ate. This allows us to understand his later work on the pure past as second synthesis of time and on destiny and freedom, to be covered respectively in the next two sections. The pure past will be defined as determining the form of the passing present – that it must pass, and how it must pass – but it does not determine or cause the content of any particular passing present. The pure past cannot be the cause of the present or completely determine it. Deleuze's philosophy cannot be deterministic.

An objection to Deleuze's argument helps us to understand its direction but also its difficulty. Why claim that a founding of passed presents is necessary, if recall depends not on going back to past presents, but rather on inspecting present codes and signs? There is no actual going back in time when a genetic code from earlier ages still operates in present organisms. There is no going back in time when an archaeologist unearths forgotten settlements, or when a pathologist reads signs of earlier violence on a shattered skull. There is no going back in time in scree shaped by a long-gone freezing and thawing of rocks. All such reference points are inspected in the present and have no reverse effects on the past. So the model of memory taken by Deleuze might be a misleading one, if we agree that memory is not a container of representations of the past that we can access and use to go back into the past. If memory is a *present* store of codes, presumably laid out in the past, but operating in the present, it looks like a very bad candidate for the function of determining past presents, because memory *causes* the current present to change into the future but does nothing to the past. This is also shown by the monastery example: they only make the past by keeping it for the present in the present. It is matter of current record, rather than a process relating a present to something into which it is supposed to pass.

The answer to this objection draws on two aspects of Deleuze's work on time. First, the second synthesis of time is, like the first, a passive synthesis. The objection turns on appeals to active cases of memory or the use of recorded code, where active does not necessarily mean deployed by an active subject, but rather a process passing from a larger set of particulars to a smaller representative set; for instance, in the way a record of shades of colours might consist of fewer types than the many variations actually encountered, or in the way a technical museum collects only the most representative machines from earlier periods. For Deleuze, however, the passive synthesis making the passing presents appropriate does not pass to representations, nor does it pass to a limited number taken from within a greater set. Instead, the passive synthesis of the past is a synthesis not of particulars but of levels, that is, not of the passing presents themselves, but of the conditions for any ordering of them. For all passing presents to be ordered and related the successive levels created by their passing have to be connected or, to use Deleuze's term, they have to be encased in one another. This brings us to the second dominant aspect of his work on time operating in the second synthesis. His argument for the second synthesis is a transcendental deduction. He is deducing the conditions for an empirical process in relation to necessary principle. Here, this means deducing the synthesis required to determine how active memory through representation is possible given the passing of presents.

So though the first moves towards the second synthesis of time appear rhetorical, they lead into a careful deduction. The helpful clue for following this deduction lies in Deleuze's use of 'Memory' as capitalised for the process of the second synthesis and 'memory' for active memory: 'Habit is the originary synthesis of time, constituting the life of the passing present; Memory is the fundamental synthesis of time, constituting the being of the past (making the present pass)' (DRf, 109). The being of the past is not the representations, records or codes of an active memory in the present. It is the condition of possibility for all the different active memories, their differences, but also their connections, above all their connections with the passing presents that came before them – all of them. Note in passing how the English translation for '*faire passer*' in the last quoted passage can be misleading in this context. '*Faire passer*' must not be understood as 'causes the present to pass' (DRe, 80). It is instead a making understood as a process of determination; to determine the form of all passing presents rather than causing particular ones.

According to a transcendental deduction, 'to be the condition of' is above all not to be a cause, because what is sought is a general condition for all manifestations of something (passing presents and active memories). So it is a deduction of a process different from the processes of the forms it is the condition for. This therefore leads to two important consequences repeated throughout *Difference and Repetition*. Condition and conditioned must be distinguished. This is because they interact according to different processes, not only internally, in the sense that the relations between habits are different from the relations between memories, but also between themselves, in the sense of the determining relation

The second synthesis of time

of Habit to Memory and the determining relation of Memory to Habit. Deleuze's transcendental deductions do not lead to separate realms. They lead to distinctions between processes such that they condition one another asymmetrically (or do not determine one another in the same way, or according to one and the same relation – such as cause, for instance). Causal processes are symmetrical in the sense that they can be reversed and run in exactly the same way: from cause to effect, or from effect to cause, the pattern will be the same. This explains the difficulty for causal accounts in terms of explaining the arrow of time, since there is no reason to assign an arrow from past to future. The whole pattern of causes and effects is fully determined and as such is neither asymmetrical nor placed in time, if time is considered to be irreversible. For Deleuze, as we have seen in the previous chapter, the arrow follows from an asymmetry in the living present. A different asymmetry governs the relations of Habit to Memory, of the first synthesis of time to the second. It is internal to the first synthesis that time must necessarily go from past to future. Once further asymmetrical processes are taken account of, we shall see many further consequences, such as the impossibility of a satisfactory representation of the past and the impossibility of a complete obliteration of anything that has been present.

THE DEDUCTION OF THE PURE PAST

Deleuze's deduction of the pure past as the second synthesis of time begins with a distinction drawn on the basis of an opposition between particular and general. As we saw in the previous chapter, he used a similar move when explaining the distinction between the syntheses of the past and of the future in the living present, where the former retained particulars and the latter expected generalities. His argument for the second synthesis reverses this order by distinguishing the past as that which a present passes into, from the past as that which is synthesised in the present. The past for passing presents is general and not particular, because it is a condition for any passing present which can then be aimed at and represented in active memory. Thus the aiming present as active memory is now particular, since it approaches the past in a particular way and for a particular aim, whereas the past as condition for any possible past present that could then be aimed at is general: 'By contrast from the point of view of the reproduction of memory, it is the past (as mediation of presents) that has become general, and the present (former as well as present) that has become particular' (DRf, 109). The past as mediation of present is the condition for their relations, for instance, the aiming at a past present from an actual one. In terms of the objection to the need for a second synthesis, the problematic move here is the distinction between particular and general. Why claim that the past is general rather than a collection of particulars? Thus, for instance, when selecting from a set of past elements according to a present code in order to build or evolve a new being, we could say that the past elements are in no way general; they are a very precise particular set, carried in the present, that we need to select within according to a code and perhaps a further set of environmental constraints (*What needs to be taken from this list of elements for a successful being to be constructed in this environment?*).

Deleuze first answers these objections by looking at active memory in the human mind as described by Hume's associationism. The point is not to build an argument on the association of ideas through resemblance and contiguity such that an active memory searches through the past for things that are either close to or resemble current ones because that is the only way any ideas can be associated.[8] Instead, it is to demonstrate the limits of active memory as representation in terms of what it can pick out from the past. Active memory, as association, only picks out artificial signs. As shown in the previous chapter on the first synthesis of time, these signs are not about the synthesis of the past in the living present, but instead depend on a distinction drawn between past and present such that something in the present is taken to represent something in the past. There is therefore a disconnection in operations of active memory between the representation and what it must pick out from the past. Active memory is representational in its mode of operation because unlike a process of passive retention, where a whole series is drawn together in one stretch or duration, active memory is a divided process: there is an identification of a set of current requirements which are then taken into the past. Therefore a representation is necessarily taken into the past.

After an argument with your work colleagues, you slam shut the door to your lab and storm off, only to realise that you forgot to bring the controversial experiment results with you. Standing outside the lab, you try to remember where you put them, so you can sneak back in and out as quickly as possible, avoiding further recriminations. Before opening the door again, you have to remember exactly where the research folder was placed. You represent it to yourself and its place in the lab. It might be there, it might not; you cannot be sure. So the lab – the past – becomes a general locus,

The second synthesis of time

not fully grasped or represented, where you can position a representation in relation to an actual representation of yourself and of the missing dossier. The human focus of such an example is not relevant, since a machine searching through a database involves the same separation when the searched-for item is first input prior to its identification in the database. At that point the item is held in relation to a general database, that is, one where its presence in the database is not yet given as an output. Before it is confirmed, that position is a representation of a searched-for item and a general database where it is a possible item. Once it is confirmed it remains a representation in relation to the general database associated with the first enquiry though it is assigned a particular place in the output. Deleuze's answer to the objection that the past is a set of particular pasts is therefore that this is not the case in relation to an active process setting off into the past in search of something. The search, the aiming towards the past, is directed at a general past. His next move towards the second synthesis of time will then be to deduce the form of this general past with greater precision.

When a past present is actually represented, this representation also includes a representation of that actual representation to itself. This is because once a past present has been selected in the general past through memory it is also positioned in relation to the aim or search in the present that set out to find it. For instance, when you run through your memories of the inside of the lab and settle on a particular location for the folder of data, you presuppose the first state where you actually did not remember where the folder was. You can reflect on this state as well as on the later search. Similarly, the output screen responding to a database request presupposes and has to relate to the initial request for it to make any sense. Otherwise, we would have to cope with Dadaist computers, randomly announcing search results with no initial request or source (*The answer might well be '342' but what was the question?*). Or searches between two levels of machine memory would break down at the point where the searched memory returned to the initial one which had not recorded the object of the search in some way.

So any active memory increases the past by a level, that is, it makes the initial enquiry into the past become an indexed part of the past. However, the status of that initial moment is different from all others within the represented memory. The passing actual present that set out to recollect has a different status to the ones it set to search within: 'The actual present is not treated like the future object of a memory, but as that which is reflected at the same time as it forms the remembrance of the former present' (DRf, 109–10). This leads Deleuze to conclude – in a very rapid step – that two processes are at work in the active synthesis of memory: reproduction and reflection. This is comprehensible though when we look at the two distinct ways in which the presents are represented. When it is a particular former present represented as the aim of a search, we have recollection and memory. When the present that embarked on the search is represented we have reflection, because it is represented to itself with the added element of the memory. The moment of reflection is also one of understanding, for instance, because it allows for an understanding of whether the searched-for item has been successfully identified or not.

Active memory therefore implies successive levels in the past, as each reflexive moment becomes past and indexes another layer: $((((\text{Past} + R') + R'') + R''') + \ldots)$. These layers contain one another and Deleuze searches for the condition of this property of containment: 'The whole problem is: under what condition?' (DRf, 110). On first reading, it is not obvious why he needs to search for a further condition here, since he has already explained how the levels are constituted and indexed on active memory. However, the key to this is that the study of the particular example of active memory is but a case of a wider property that needs to be explained in terms of its formal conditions: how is it possible for any past present to be reproduced in the actual present? Here is Deleuze's answer: 'A former present is reproducible and an actual present can be reflected through the pure element of the past, as past in general, as *a priori* past' (DRf, 110).

For any former present and any possible actual present to be reproducible, the past cannot be a particular collection of former presents, since this could discount some further possibilities. It must instead be a general past, that is, one that allows for any reproduction. It must therefore not be a past dependent on a particular experience of presents, a subset of occurrences, and must hence be *a priori* (prior to any given experience). Furthermore, since it is not a collection of particulars, it is pure, in the sense of not characterised by or limited by any particular set. Deleuze frequently uses this special meaning of pure in *Difference and Repetition*, where pure means free of actual identities, of particular beings and of any representations. The conditioning process ('another (transcendental) passive synthesis proper to memory itself') does not involve the same processes as the conditioned: 'the (empirical) passive synthesis of habit' (DRf, 110).

The second synthesis of time

Deleuze then develops the idea of the pure past on the basis of a reading of Bergson's *Matter and Memory*. This should not be seen as implying that Deleuze's concepts and arguments are the same as Bergson's. On the contrary, they are developments of them. This is an important point for the study of Deleuze's work on cinema in relation to his work in *Difference and Repetition*, since as we shall see in Chapter 7 Deleuze's cinema books are closer to Bergson's version than his earlier interpretation. The explicit aim in *Difference and Repetition* is to explain exactly what the pure past is, in relation once again to a series of paradoxes. The pattern of this construction on the basis of critical and productive paradoxes is instructive in terms of understanding the general traits of Deleuze's philosophy of time, since we might have expected a philosophy to seek to avoid such paradoxes, or reject hypothetical theories of time because they lead to such paradoxes. In fact, one of the distinctive features of Deleuze's philosophy of time is to embrace paradoxes for their productive power. This is related to the power ascribed to problems in *Difference and Repetition*; like problems, paradoxes cannot be resolved but must rather be transformed creatively within a necessarily speculative model. It could be said that paradoxes prepare the way for problems through a critical clearing of the commonsense certainties of a field and through the generation of a structure of opposed, yet connected and irreducible principles. Deleuze identifies three paradoxes relevant to the pure past in *Matter and Memory*. These are: the past must be contemporaneous with the present that it was; all the past must coexist with the new present in relation to which it is past; and the pure element of the past pre-exists the passing present. Rephrased in more simple terms, the first three paradoxes are: since the past adds nothing to the present that passes into it, it must be contemporaneous with that present; since the past must be contemporaneous with each passing present, all the past is contemporaneous with each passing present; and since all the past is contemporaneous with each passing present, the past is contemporary with all of time and pre-exists any passing present.[9]

In order to understand the way these paradoxes work in Deleuze's argument, it is helpful to separate their critical and productive functions. Critically, they work against the idea of the past as a collection of particular past presents and hence against the idea of active memory as sufficiently determined by such an idea, as the power to recall such past presents. They therefore also support the transcendental deduction of another version of the past, the pure past, not resembling such a collection. For instance, the past must be contemporaneous with the present since otherwise the past would have to be different from the present it was. If we could then take time travellers back in time, they would necessarily find situations different from those they had experienced in the present (*I was not standing there!*): 'We cannot believe that the past is constituted after it has been present, nor when a new present appears' (DRf, 111). Yet, this is difficult because then we have no way of explaining how a present passes if there is no difference between the past and the present that it was. As soon as the time travellers landed back in time, they would be completely back where they were and time would begin at that point, necessitating an erasure of their memories, disappearance of the time machine and so on: 'If the past had waited for a new present in order to constitute itself as past, the former present would never pass nor the new one arrive' (DRf, 111). This is because the new present can change nothing in the former one.

The move to the pure past in this first paradox is then to set the past within the passing present contemporaneous with it but not identical to it. However, though the past and the passing present are contemporaneous, they have different forms. So the creative move is to replace the idea that the past is the same as the present that was, with the idea that the past is a different kind of condition for the passing of the present occurring with the present or contemporaneously: 'A present would never pass, if it was not past "at the same time" as it was present; a present would never be constituted, if it was not first constituted "at the same time" that it was present' (DRf, 111). Deleuze has therefore replaced a notion of simultaneity where two things of the same kind are simultaneous, with a relation of contemporaneity between an actual present that can be represented (as present or past) and a different element, the pure past, accompanying every present and making it pass. In deducing the pure past as different in form from the present, Deleuze avoids the paradoxes generated by the identity of the past and present. This, however, does not tell us much about the pure past, about how it makes the present pass, or what it is. So he proceeds to the next paradox, about the coexistence of all of the past with any present, in order to show further aspects of the pure past.

This second paradox depends on the same assumptions as the first. If there is no difference between the past and the present that it was, then the whole of the past must accompany each new present, since otherwise we would have a way of distinguishing the past and the present on the basis of the new present. If we travel

The second synthesis of time

back in time at different yet close times to different pasts, then we will have to explain how those pasts became different and we cannot do so without contradiction of the identity of the past and the present that it was (*Last time we came back to this time, the sky was grey and the phone box was bigger. How can that be?*). So it is the same time and the whole of time that accompany each new present. Yet this whole of time cannot be a collection of representations, as has already been shown. So now we know that the pure past is the whole of the past and cannot change in relation to each new present in the way a collection of copies of presents might: 'That's why, far from being a dimension of time, the past is the synthesis of the whole of time and the present and future are only its dimensions' (DRf, 111). Here, the relations of times in terms of dimensions have changed from the first synthesis of time, where the past and future were dimensions of a contraction in the present. In the pure past, the present becomes a dimension of the past.

The importance of the idea of dimensions is two-fold: it stresses how in each synthesis there is only one time, respectively the present and the past, for the first and second synthesis; it also specifies different processes within that unique time, allowing different dimensions to be defined. This is where Deleuze's path into the second synthesis is necessarily somewhat misleading since, because it begins with the active memory and the search for the condition for the passing present, it leads us to retain the present as independent of the past within the second synthesis. This is not at all the case. In the second synthesis, the present is the most contracted state of the passive synthesis of all of the past. It is no longer a synthesis of a particular pattern from the past in the present, but rather a dimension of an ongoing synthesis of all of the past in the past. We therefore have two sides of any present (as we shall see, there will be another with the third synthesis of time). There is the present as contracted synthesis, a particular stretch in the present, and there is the present as the most contracted state of the all of the past, of the pure past. Neither of these times can be reduced to one another and Deleuze's philosophy of time is therefore one where time is only complete when taken from different sides or perspectives: a time of the living present and a time of passing present in relation to the pure past.

The present as first synthesis of time is necessarily accompanied by a synthesis of all of the past. This means that any present determined as a limited stretch or contraction must pass, because the past as synthesis is an ongoing process of becoming that determines every such present as passing. In turn, this provides the response, in terms of the pure past, to the paradox stating that the past must precede the present that it makes pass or engulfs. It precedes it because the present is only a dimension of the pure past that must therefore pre-exist it. The pure past is synthesised before each present:

> The paradox of pre-existence therefore completes the other two: each past is contemporary to the present that it was, the whole of the past coexists with the present in relation to which it is past, but the pure element of the past in general pre-exists the passing present.
>
> (DRf, 111)

We must pay very close attention to the choice of vocabulary here. There is a shift through this quoted passage along different relations of past and present. The past is *contemporary* to a present that *has past*. Contemporaneity is posited on a present that is no longer a living present. The whole of the past *coexists* with a present *in relation to which it is past*. Coexistence is posited on a present that can return to it in active memory. It is the condition for such activity; active memory could not return to the past unless they coexisted. The past in general *pre-exists* the *passing present*. Pre-existence is in relation to a present that is made to pass. Pre-existence is the condition for the passing present. *A rotten apple falls. When fallen it is contemporary to the whole of the past. As it is falling, it coexists with the pure past; both changing with the fall. The apple can only become past, though, because the past pre-exists it.*

However, when viewed from a present independent of the past, the paradox of pre-existence remains. How could the past precede the present it is the past of? It must do so, if the whole of the past comes before each new present. In the second synthesis of time, the importance of the statement that the present becomes the most contracted dimension of the past lies in the reversal of the roles of maker. When the past is the dimension, as in the living present studied in the previous chapter, events in the present make the past as a synthesis in the present. On the other hand, when the present is the dimension of the pure past, the present becomes something made. It is not made in the sense of the creation of particular characteristics, but rather in terms of essential properties, the main one of which is that every present must pass and is accompanied by the pure past. According to Bergson's metaphor, as reported by Deleuze, the present is the tip of an infinite cone which stands for the pure past:

The second synthesis of time

> The present is only the most contracted degree of the past coexisting with it, if the past first coexists with itself at an infinity of degrees of relaxation and diverse contraction, at an infinity of levels (therein lies the sense of the famous Bergsonian metaphor of the cone, or fourth paradox of the past).
>
> (DRf, 112)

According to the metaphor, the present must pass because the cone is itself a process of becoming that is misunderstood if we take an identification or representation of it or of the present as true representations of the cone. Such representations are necessary yet always miss that which they try to represent: 'But truth is that the general idea always escapes us, as soon as we attempt to fix one or other of these two extremities' (Bergson, 1959: 302).

These last lines are Bergson's and it is instructive to read his versions of the situation of the present as a dimension of the past, since they are more dramatic and less formal and paradoxical than Deleuze's: 'Practically, we only perceive the past, the pure present being the ungraspable progress of the past gnawing at the present' (Bergson, 1959: 291). Bergson's much greater emphasis on observation distances the two treatments of memory. He studies memory through a close analysis of consciousness and in terms of a critique of various accounts of mind, in particular associationism. Bergson's argument does not strictly identify paradoxes, but rather demonstrates incoherence in representational, content-driven and spatial accounts of consciousness and of time by showing how they lead to contradictions. Each reductio ad absurdum is then shown to be avoided if we give a different account of the workings of consciousness, such as the claim that 'we only perceive the past', that is, we only perceive a duration or stretch of time limited by an empty pure present.

This contrast with Bergson draws out a number of features of Deleuze's mode of argument and suggests a further series of critical points to be made against it. The following passage from *Matter and Memory* involves a different use of the concept of condition to the transcendental one at work in *Difference and Repetition*:

> Our entire past psychological life conditions our present state, without determining it in a necessary manner; it also reveals itself in entirety in our character, although none of the past states is manifested explicitly in the character. Together, these two conditions ensure a real, though unconscious, existence to each of the past psychological states.
>
> (Bergson, 1959: 289)

We can see here that 'condition' has a different meaning to Deleuze's. Bergson is not deducing a general transcendental condition for a formal process (such as the passing away of the present). Instead, like Deleuze, he is offering an alternative to the concept of cause, but unlike Deleuze, he is doing so in order to give an account of how each individual consciousness relates to its past as shown in the true operation of memory. This is where we can raise the question of the legitimacy of Deleuze's work when compared with Bergson's. What is the validity of an account of the past that does not base itself on a scientific account of causality (or some other contemporary candidate for explaining relations between states of affairs scientifically) but equally does not observe the operations of memory in detail or offer a full theory of memory in relation to consciousness, but instead constructs a speculative transcendental frame with abstract terms such as the pure past?

DESTINY AND FREEDOM

One of the answers to an objection to Deleuze on the grounds that he does not observe memory or consciousness in a thorough or consistent way combines two features of his work in *Difference and Repetition*. He is not primarily concerned with human memory or consciousness, but rather in a general study of repetition in relation to time. He is not constructing a philosophy according to an empirical approach, but rather combining a minimal observational element with a series of transcendental deductions guided by a speculative conceptual frame. These features are brought into his work on time immediately after his study of the role of paradoxes in the deduction of the pure past as the second synthesis of time.

First, Deleuze draws a distinction in the way repetition works in the first and second syntheses and therefore in the passive syntheses of habit and of memory. The difference is important because it develops the idea of repetition much further in the direction of his new conception of its form. This is made possible by focusing on the difference between a repetition based on the succession of elements (as explained in the previous chapter) and a repetition understood as degrees of contraction of a whole 'that is in itself a coexistent totality' (DRf, 112). The distinction is more easily approached through an analogy. If you take a shelf of books by your favourite authors and add the latest one to be published, you contract a series of elements in a novel manner, for instance, in drawing out an unforeseen comic element in all of them. This would be an analogy for the first synthesis, where the past is a dimension of the present as contraction. However, if you take all the past degrees of emotions that could potentially be reawakened in a new reading of any book, you have a totality in itself. Nonetheless, this totality can be contracted differently in terms of the degrees of its internal relations in relation to different circumstances. For instance, some relations can become more important, others less so. When the present contraction, with its novel comic element, is made to pass into the whole it constitutes a new level for it where all the degrees are present but in a novel set of relations. There is no repetition of elements in this whole, nor can it be organised into a sequence. It does though allow for variations in degree where the most contracted variation, the one at the highest level, is defined as the present.

Second, Deleuze applies his work on levels and dimensions of time to the example of 'a spiritual life' in order to show its implications for the way we understand repetition, destiny, determinism and freedom in such a life (DRf, 113). The choice of words here is somewhat awkward, since contemporary definitions of 'spiritual' associate the term primarily with the idea of a religious life, a life of the soul, or a life of devotion. The French term '*esprit*', though it can mean 'soul', also has a more neutral sense as 'mind', mainly if this latter is understood as disembodied. It is thus closer to 'spirit' in its non-religious sense. The 'spiritual life' he turns to at this point of the book is therefore best understood as the life of a mind or spirit that has a consistency through time; it is the life we indicate in expressions such as 'I have lived my life to the full' or 'My life has always been governed by curiosity'. It would be a mistake, therefore, to assume Deleuze is talking about a mystical life or taking such a life as the paradigm of all lives. Though Deleuze is working from a basis in Bergson, his method cannot be associated with any purported Bergsonian mysticism.[10] Instead, the point of Deleuze's reference to the life of mind is to draw attention to our reflection back upon a life as ours – the life of a mind – rather than a predetermined causal series of material facts.

Deleuze's initial move is to consider a common definition of destiny that he will then alter in relation to the first synthesis of time and in relation to this notion of a spiritual life: 'Nevertheless, we have the impression that, however strong the possible incoherence or opposition of successive presents, each one plays "the same life" at a different level. This is what we call destiny' (DRf, 113). The living present is multiple and consists of many overlapping durations and stretches of different lengths. We have relatively short durations: *the time to finish this coffee.* We have, relatively, very long ones: *my life as a daughter; my life as father.* They have possible oppositions and incoherence: *his life as a queen's council and his life as a republican traitor and his life as expert on the baroque and his life as a mathematician.* This preliminary definition of destiny is explicitly loose and Deleuze is careful to point out that it is based on an 'impression', that the incoherence and opposition are 'possible'. He is also cautious in setting the same life in scare quotes. This is for two reasons: he will show later that this life cannot be considered the same through time or indeed at any time; he will also show that this life cannot be considered as someone's life, or a particular life. Much later than *Difference and Repetition*, Deleuze returns to the problem of how to convey this novel idea of life in his last essay, 'Immanence: a life . . .' There, the scare quotes are replaced by the much more sophisticated usage of 'a life . . .' (*une vie* . . .) separating it more strongly from associations with individual persons and with a continuous and represented identity.[11]

From the starting point of this common definition, Deleuze then goes on to construct a much more sophisticated version. First, he opposes destiny and determinism. Destiny is never to be determined, in the sense of causally determined or, more simply, in the sense of a predetermined order of successive presents. There is instead a place for freedom in this novel understanding of destiny. The initial reason for this can be found in the multiplicity of the first synthesis of time as it relates to lives.[12] As we have seen, the first synthesis implies overlapping durations or stretches that cannot be reduced to a single line, or to a dominant narrative. The traditional conception of a life as a single continuous time line in a continuous well-ordered space is therefore replaced by a fragmentary life, with jumps, returns, gaps and resonances resistant to a satisfactory situation on a single continuous line: 'Between successive presents, it implies non-localisable links, action at a distance, systems of replay, of resonances and of echoes, objective chance, signals and signs, roles transcending spatial situations and temporal successions' (DRf, 113).

The key to understanding these broad claims is that Deleuze moves from the observation of the different levels implied by this irreducible multiplicity of durations to the condition for their participation in the 'same life'. Without such a move he would be open to the criticism that his position is straightforwardly contradictory because it combines a claim about an irreducible fragmentation with a claim about unity. How can we call the fragments 'the same

The second synthesis of time

life' if they cannot be considered as belonging to an identical time line and space associated with that life?[13]

Deleuze's answer avoids this problem by not looking at the overall identity of a life, but instead moving even further away from this sense of actual identity and towards the second synthesis. The condition for connecting fragmentary durations into a life is that they are playing the same life but at different degrees and levels. This reference is to the degrees of contraction of the pure past, and the condition for the connection of different durations is that they are all conditioned by the pure past because each one is a passing present made to pass by the pure past and existing as contraction of the pure past:

> The succession of actual presents is only a manifestation of something deeper: the manner in which each one retakes the whole life but at a different level or degree than the preceding one, all levels and degrees open to our choice from the bottom of a past that was never present.
> (DRf, 113)

I have translated '*reprend*' by 'retakes' in this passage, rather than the original translation's 'continues'. This is because each present replays the whole life and transforms the earlier ones, a sense we find in the English word 'reprise' and a conception of a repeat of a musical work that is very important to Deleuze: 'Each one chooses its pitch or its tone, perhaps its words, but the tune is indeed the same, and beneath all the words a same tra-la-la in all possible tones and all pitches' (DRf, 113–14; DRe, 83–4). Continuation keeps too much of succession and sequence and leads to a contradiction with the idea of taking on the whole life, rather than simply continuing on from the preceding moments. The meaning of freedom in relation to destiny in Deleuze is then not the freedom to add to a sequence, for instance, when a new director adds a new film to an established franchise (*My Life IV*). Instead, we are free to make a new cut of an existing film (*My Life, The Director's Cut*).

The analogy with making a new cut of a film, or of staging a new version of a play, draws out many of the oddities and difficulties of Deleuze's ideas on freedom and destiny. In his model, you can make a new cut, but no reel or even frame can be left out, no new scene can be added and no old one can be shot again. This is because we replay the pure past in relation to all the passing presents and, as we have seen, the pure past is all of the past. This explains why we have to make a reprise of all of the past, as stated in the passage quoted above. It also explains Deleuze's at first sight strange statement that this is a past 'that was never present'. The past we have to retake is pure; it contains no particulars or actual presents. What is more, this pure past can only be replayed in terms of its degrees and levels, that is, in terms of degrees of relations between levels. Retaking the past is therefore never changing actual events that have happened. It is rather to select the intensity of degrees we assign to different levels which then operate in the passing of all presents. This in turn means that the value and significance of those passing presents vary with the changes in degrees. A present that appeared to be at a high level and dominant degree can find itself at a lower one. A present that appeared to be both insignificant and distant in time can grow in importance and find itself at the centre of a nub of intense relations. Drawing back to the film analogy, though a new cut retains all of the scenes from the earlier versions, it can change their sense, value and emotional significance through a novel ordering (*put death in the middle and birth at each end*). There are risks in such examples, though, because for Deleuze's philosophy oddity is really in analogical thinking rather than in the creation of novel concepts such as the pure past. We can only ever repeat all the past, not as representations, such as frames from a film, but as relations of level and degree. This shows the inherent danger of analogy for transcendental philosophy. Despite its advantages in explanation, analogy has too strong a dependence on representation and symmetry to fully express differences in realm and relations so important for transcendental deductions. So long as our pedagogical culture depends on analogy and representation, the explanation of the validity and form of the transcendental will remain difficult and prone to misunderstandings.

Freedom exists in relation to destiny and determinism for Deleuze because we are free to change the relations of level and degree given to all past events through our present acts. We are not free to change determined relations between actual presents. Here, changes in level and degree can be understood as changes in the intensities of distinctness and obscurity of relations in the pure past, that is, some relations in the past will be made more distinct as others become more obscure. For example, an act of atonement in the present can change nothing of the actual acts it seeks to atone for. It is free, though, to change the hold such acts have on new passing presents, perhaps by making them less significant in their relations to other events, or by making them more obscure and distant, and thereby diluting their hold on novel ones. Thus, to heap betrayal upon betrayal might increase the intensity of treachery as a line leading from the past to the present, whereas to forgive might weaken it. Within Deleuze's metaphysics, this freedom exists because the pure past makes all presents pass and coexists with them. When our acts are made to pass they too interact with the pure past and thereby interact with all passing presents. The reason this does not contradict causality lies in the distinction he makes between the empirical character of presents, 'their associations according to causality, contiguity, resemblance and even opposition', and their noumenal character, the relations of 'virtual coexistence between levels of a pure past' (DRf, 113).

Deleuze picks up on Kant's concept of the noumenal and Bergson's idea of the virtual in order to distinguish the second synthesis of time from actual presents. The pure past is noumenal; it is a condition for the passing of actual passing presents, but it is not itself actual. The pure past is virtual, that is, ideal but not in the sense of ideas in active consciousness, but rather in the sense of relations between levels in the pure past. It is worth noting that this combination of Bergson and Kant is highly original and surprising, given Bergson's critique of Kant's transcendental philosophy.[14] However, the combination works because Deleuze departs from Bergson's work on consciousness and human memory and from Kant's description of the noumenal as the realm of things in themselves beyond our understanding. He therefore arrives at a new philosophical position with a transcendental pure past (the second synthesis of time) in a relation of reciprocal determination with the present (as first synthesis of time). The noumenal then becomes a realm that all actual things determine and are determined by. It becomes a virtual and ideal realm as condition for all events and not just those of human memory.

This deduction of the second synthesis of time is then concluded on two important but very different remarks. One is practical and concerns an application of the second synthesis as synthesis of the whole of the past. The other is speculative and explains the role of difference in the diverse syntheses and repetitions of time. If misread, the application of the second synthesis of time can lead to another interpretation of Deleuze as a mystical philosopher. This is because he seeks to explain metempsychosis, reincarnation or the transmigration of souls, on the basis of his account of time. Given that any passing present is the most concentrated state of the synthesis of the whole of the past, each present is a reprise of all the lives that preceded it:

> Since each one is a passing present, a life can take another on, at a different level: as if the philosopher and the pig, the criminal and the saint played the same past, at different levels of a giant cone. This is what is called metempsychosis.
>
> (DRf, 113)

The reference to the philosopher and the pig might be Deleuze's humorous reply to Mill's famous and often misquoted dictum on human beings, pigs, Socrates and fools: 'It is better to be a human being dissatisfied than a pig satisfied; better to be Socrates dissatisfied than a fool satisfied' (Mill, 2002: 12). Since in the pure past all lives replay each other at different levels, the human is in the pig and the philosopher in the fool. More deeply, for Deleuze, humans are not fully human until they express the pig within them and the true philosopher is one who is also or even foremost a fool.[15] The reference to the criminal and saint might be an echo of Sartre's book on Jean Genet, *Saint Genet: comédien et martyr*.[16] This is not mysticism, though, and it is important not to miss the 'as if' in Deleuze's sentence. The pervasive idea of transmigration can be explained because each life communicates to all others as actual presents through the shared medium of the pure past. It is only as if we replay actual lives though, because we really replay the levels and degrees of the pure past, which contains nothing actual. The use of real in opposition to actual is important here (and throughout *Difference and Repetition*). The transcendental virtual is real and not imaginary or abstract. It is real because it completes the actual in real processes of reciprocal determination. Deleuze's philosophy is radically inconsistent with actual reincarnation, since every actual present passes and can never return. It is, however, consistent with a novel and very difficult notion of return through the return of difference in relation to the pure past.

This return of difference is the second important remark made after Deleuze's work on repetition in the second synthesis of time. In material repetition, the synthesis of the living present or first synthesis of time, difference is subtracted, because a selection is made of a particular series within many differences. In 'spiritual repetition', the second synthesis of time, difference is included, because all differences are taken up, but at a particular level and degree. What is important, though, is that these different presents and syntheses belong together and complete one another:

> The present is always contracted difference; but in one case it contracts indifferent instants; in the other, by passing to the limit, it contracts a

differential level of the whole that is itself relaxation and contraction. This is such that the difference of the presents themselves is between the two repetitions, the one of the elementary instants that it is withdrawn from, and the one of the levels of the whole in which difference is included.

(DRf, 114; DRe, 84)

Note how this study of difference in relation to repetition is a direct echo of the closing remarks on the first synthesis of time. Difference is not an intrinsic property or essence of any repetition. It is rather the reason why each repetition is conditioned by another, where the condition explains how a repetition of the same elements requires the adding in of a variation to that same repetition and how the repetitions of pure variations or differences requires a subtraction of difference in order to be determined.

Note also that this role of difference in itself in both repetitions is also the reason why neither repetition can be represented, because the difference they either subtract or add cannot be represented, since it is always between two realms. When repeated elements are represented the subtraction that representation depends upon is erased. When repetition within the pure past is represented a subtraction is imposed on it such that it is no longer the whole of the past. Though Deleuze's argument and structure can seem loose, in fact, they are highly rigorous and there is a careful pattern to his deductions: beginning each time with a cursory observation; deducing transcendental conditions revealing syntheses dependent on repetition; then reflecting on the principles emerging with these syntheses; applying these back in a novel way to actual cases (such as metempsychosis); then finally drawing all the syntheses together through an explanation of the role of difference in itself in repetition. It is because difference is between the living present and the passing present that they belong together. One subtracts from the other while the other adds it back, but always differently in an ongoing creative process.

HOW TO SAVE ALL THE PAST FOR US?

Through this study of the second synthesis of time we have seen how it provides a response to a series of objections. It explains how there can be an account of the whole of the past and a conception of a complete account of time without reducing time to a set of disparate elements or to a hermetic and fully determined whole. We have also seen how Deleuze avoids having to posit nothingness or a void in his account of time, yet does not fall on to an account of eternal beings. Things pass, yet they do not pass into nothingness. We have also seen how 'Memory' or the pure past is a foundation for time without being an unchanging ground. On the contrary, the past is in continuous flux and therefore induces each present to pass. However, this raises a practical question that leads into a practical problem. How should we act, given this relation between past and present? Does the impossibility of a full representation of the pure past absolve the present and in particular the present as active memory of any responsibility to the past?

Deleuze's answer to these questions will lead him to consider a third synthesis of time. He leads us into this conclusion through the presentation of a practical problem, where practical can be understood as meaning a moral problem or problem of moral action. Moral, here, should not be understood as connected to specific moral dilemmas. His work is not primarily concerned with questions of the right action in this or that social situation. Instead, it is to reflect on some of the properties of the second synthesis of time in relation to action. Can we penetrate into the pure past, even though we cannot represent it? Can we live with the pure past as we do the living present, as an ongoing synthesis we can learn to accomplish through an apprenticeship to signs (as studied in the previous chapter, here)? Deleuze follows these two reflections with a further one requiring much more interpretation, since it introduces new concepts that have not yet featured in his deductions of the synthesis of time or in his speculative framework:

> The whole past is conserved in itself, but how to save it for us, how to penetrate into that in itself without reducing it to the old present that it was, or to the actual present in relation to which it is past. How to save it *for us*?
>
> (DRf, 115)

I have given the French '*pour nous*' as 'for us' here, rather than the original translation of 'for ourselves' (which would have been given as '*nous-mêmes*'). This is to avoid the idea that Deleuze is concerned with the quite traditional idea of preserving the past in memory as a benefit to the self, so it may know itself through its past.

The way Deleuze has introduced the problem tends to the idea that it is a question of how to live with the past. He has introduced two new terms in his study of time, two rare terms in *Difference and Repetition*: 'save' and 'us'. The emphasis on the italicised us is important, as is the choice of 'to save', which is further from conserve, or preserve, or keep than in English. It has a stronger sense of salvation, peril, preserve (in the sense of preserve from something) and avoidance of a specific loss (loss at sea for instance). This is a dramatic moment in Deleuze's book and treatment of time. The second synthesis of time leads to a problem of representation, which itself leads to a problem of how to live with the past. These then lead into the question of how to save the past, to save it from loss, *for us*, that is not for itself but for us, for our living present. The past needs to be saved from oblivion for the benefit of the living, or more precisely, for living beings together.

This is also a dramatic shift into a moral problem with its characteristic plurality (it is a problem of togetherness) and its characteristic difficulty (things are at peril in a way that is difficult to resolve). The shift is accompanied by a change in lead thinkers. Deleuze moves from Bergson to Proust, a figure he has turned to often in *Difference and Repetition*, harking back to Deleuze's earlier book on Proust, *Proust and Signs*. Though Bergson has shown the limits of representation in relation to virtual memory through his critique of consciousness and of active memory, he has not shown the way to live with the past, given the failure of representation – at least in *Matter and Memory*. Proust, however, through his study of reminiscence, has shown how the past can be saved for us without reducing it to representations of a former age or to representations of our age (the past as how it could be). Instead, the past is given 'as it was never lived, as a pure past revealing its double irreducibility to the present that it was, but also to the actual present that it could be, in favour of a telescoping of the two' (DRf, 115). Reminiscence shows that the past is lost and forgotten as a past present. It accepts it. Nonetheless, it still saves the past by taking the past representation and the present representation and, without making them the same, it makes a third image with them. It is in this special kind of forgetting and recreating that the past is lived with in forgetting. In the more technical treatment we have been following, the condition for this telescoping of the two images is the pure past. It makes both presents pass, contemporary to them and pre-existing them.

The answer to Deleuze's questions therefore lies in his understanding of the process of the second synthesis of time as pure past. It is because the pure past is a process on all passing presents that it can be saved for us. It is because it cannot be represented that it must be saved as forgotten, that is, as recreated in the present. But it is also because it pre-exists all presents that a pure forgetting, an obliteration, cannot save the past. Only as a recreation of past presents as forgotten, as in need of being lived differently, can the past be saved for us. It needs to be saved for us because our living presents are made to pass by the pure past and are pre-existed by it. Yet this is exactly the point where he detects the need for another synthesis, because though the pure past shows us *what* we must create with, it does not show us *how*: the echo of the two presents only forms a persistent question, developed in representation as the field of a problem, with the rigorous imperative of searching, answering and resolving (DRf, 115). In Proust, the answer to the 'How?' question is through Eros: it is always through an erotic attraction that we are led to an answer as to how to save the past. Deleuze says that Eros shows us how to penetrate Mnemosyne or Memory (the second synthesis of time). This is learned from Proust's signs. To answer the questions of why it is Eros, Deleuze passes to the third synthesis of time.

4

The third synthesis of time

FROM DESCARTES TO KANT

In *Difference and Repetition*, Deleuze begins the transition from the second synthesis of time to the third through a discussion of Kant in relation to Descartes. This discussion is inserted between two references to Eros or to a drive explaining a process of selection in relation to memory that must be added to the synthesis of time in memory and the living present. This additional process cannot simply be drawn from the first or the second synthesis, yet Deleuze will show its necessity for both. We shall follow his arguments for this throughout this chapter. However, prior to this analysis it is important to understand why Deleuze sets the scene with his work on Descartes and on Kant. Only then will it be possible to understand the full implications of his turn to Eros in the context of a discussion of the subject and of the self. This turn is open to many misinterpretations since it involves great transitions from our usual understanding of terms such as 'subject', 'time' and 'Eros'. It could be objected to this approach that the core of the third synthesis of time lies in Deleuze's reinterpretation of Nietzsche's doctrine of eternal return. That eternal return is essential to any conception of the third synthesis of time is not to be denied here. Eternal return will be given a chapter of its own next. However, if the Kantian antecedents to Deleuze's account are left out, the relation of the third synthesis to the second will be misunderstood. More seriously, the strength of Deleuze's arguments for the necessity of the third synthesis will have been missed.

Deleuze analyses the philosophical shift from Descartes to Kant in terms of the subject, time and determination.¹ There is a risk inherent to this discussion of the subject, since it invites the claim that Deleuze's arguments are grounded on the subject as determined by time. This then leads to the subsequent criticism that his philosophy of time is still subject-based and therefore only offers a partial and skewed account of time, running counter to his work on passive syntheses. This view is mistaken, though, because Deleuze's point is much further-reaching. Due to its determination in time as a process, the subject is fractured and this presupposes fault lines denying any priority to either a subject as philosophical foundation or to a specific form of time independent of prior processes. Thus the work on Descartes and Kant is not designed to move from the former to the latter, but rather to draw a lesson from that shift with implications taking Deleuze's philosophy beyond both. The core of the argument is then neither about the subject, not about time as such, but rather about the importance of a philosophical method and the consequences it implies for the end of the subject as philosophical foundation of any kind (logical, intuitive, phenomenological). Thanks to this method, Deleuze is able to add the idea of the death of the subject through its fracturing to the idea of the death of God, implied by Kant's work on time but not fully followed through by him.

We have already seen the importance of transcendental deductions for Deleuze's philosophy of time in the first and second syntheses. In the discussion of Descartes and Kant, the method returns in a form repeated throughout *Difference and Repetition*. This form adds a precision to the definition of transcendental deductions. They deduce the conditions for a difference across inseparable ontological realms as they connect to and determine one another, where the form of a difference in one realm is conditional on a difference in another. The precision therefore concerns the concepts of determination, indetermination, determined and determinability that are developed in relation to Kant. The transcendental method in *Difference and Repetition* explains conditions for difference in terms of determination in different modes. As we have already seen in the previous chapters, this has nothing to do with causes, nor does it seek to determine the conditions for general forms. This latter point seems rather odd when we are speaking of a shift from Descartes to Kant concerning the subject. Is not the 'I' in the proposition 'I think' either a general subject (the 'I' we all refer to when we state the proposition) or a general form (the form of the subject presupposed each time we state the proposition 'I think')?

The third synthesis of time

The whole point of Deleuze's study is to draw the 'I' away from general conceptions and back to a singular self as condition for a now singular subject stating the proposition 'I think'. In turn, this is not to ground his philosophy on that subject, nor even to claim that it is necessary to work from a study of the subject back to something else that is more fundamental – a deconstructive strategy that would still assign an essential place to the subject. Instead, Deleuze's main point is about determinability, passivity and time. The subject presupposes the self. The self passively presupposes a form of determinability. This form of determinability implies a fracturing of the self and hence of the subject because it is a pure and empty form. The argument is difficult and it is worth studying it in detail. First, Deleuze introduces the concepts of determination and indetermination in Descartes. The 'I' is determined as a thinking 'I' – it is thinking – but it remains undetermined as the 'I' that exists in the proposition 'I think, therefore I am'. How the 'I' exists is left open. Nonetheless, Kant then adds another concept that will allow him to move beyond this openness and add time to the Cartesian model. The 'I' is determinable in many ways; we can determine its mode of being in many ways. What is the condition for this? It is that the 'I' is in time. All the ways in which the 'I' can exist presuppose that the 'I' exists in time. Yet Deleuze does not want to draw a point about time and the subject here. Instead, he is concerned with two consequences. First, activity presupposes passivity. Second, this passivity involves a different form to active consciousness and one that is inaccessible to the active one.

Thus, when you actively conceive of a proposition such as 'I am breathing', you can choose to vary the attribute, from 'breathing' to 'walking', for instance. You cannot control or deny the way in which both those conceptions must take place in time; the denial itself even presupposes time. This is the passivity Deleuze is concerned with. It is a double passivity, though. This is because *it is not the subject of the active conception that is directly passive, but rather, it is passive through a self positioned in time.* The 'I' that conceives of the proposition is different from the 'me' positioned in time by being a living and sleeping thing. So now being is divided between an active subject and passive self where any action by the subject presupposes that self because the subject is only passively determinable in time through the self. A passive self is the condition for any active subject. The 'I' is therefore fractured or traversed by a fault line, because of the way the self is determinable in time:

81

> It is as if the 'I' is traversed by a fault line: it is fractured by the pure and empty form of time. In this form it is the correlate of the passive self appearing in time. This is what time signifies: a fault or fracture in the 'I', passivity in the self; and the correlation of the passive self and the fractured 'I' constitutes the transcendental discovery or the element of the Copernican revolution.
>
> (DRf, 117)

In this passage, Deleuze is changing the meaning of Kant's revolution in philosophy away from its parallel with Copernicus and the transfer from earth to sun as centre of rotation of the planets.[2] This is because Kant orientates philosophy away from empirical objects and back to the subject in time, whereas Deleuze pushes this movement further in insisting on a fractured subject with a passive self as correlate. *There is no single identified thing to rotate around, but instead a process of fracturing carrying all that rotates around it into its fault lines.* The relation of subject to self is such that they can no longer be seen as a centre at all. Where Kant re-established a centre, Deleuze sets up another paradox: 'Here commences a long and inexhaustible story: "I" is another, or the paradox of inner sense' (DRf, 116; DRe, 86).

To understand this paradox we have to return to a statement Deleuze makes twice in the space of two paragraphs but with no explanation or qualification. The presupposed form of time in the fractured 'I' is 'pure and empty' (DRf, 117). There are two questions to pose here. First, why must it be pure and empty? Second, is it indeed pure and empty? The form of time must be pure and empty since otherwise it will lead to a determination of the self, rather than an open determinability, and therefore not to inner sense as paradox, but rather as a determined and fixed relation of subject to self. Deleuze shows this through an account of why Kant fails to conclude that the self is fractured, despite responding correctly to the counter: Couldn't the subject or a proposition be situated outside time? The reason for Kant's failure is similar to the reason why Kant criticised Descartes for not situating the 'I' in time. When the proposition 'I think, therefore I am' is conceived of as outside time it is conceived as instantaneous, or as not taking any time at all. However, as such it raises the problem of continuity through time, since separate instants cannot overlap. Descartes resolves this problem by appealing to God and to the idea of His continuous creation of the world.

According to the Cartesian argument, we do not appear and disappear in an instant when we conceive of a proposition, because

God creates the continuity of our identity thanks to His identity. The unity and identity of the subject is dependent on the unity and identity of God. The difficulty is that although Kant criticises Descartes's reliance on God as secured by logical deduction of his existence, Kant reintroduces an identity for the subject and for the self, respectively through synthetic active identity (the way in which a free moral action establishes the identity of the actor) and the restriction of passivity to receptivity (so the self becomes something that is purely receptive rather than as transforming of other things even in its passivity). This pure receptivity then becomes a form of prior identity when compared with the ever-changing, reciprocal and open syntheses defended by Deleuze in his account of the first and second passive syntheses of time. In place of a fractured subject we then have a receptive self with all the conditions for that receptivity, such as restrictions on what can or cannot be received. The paradox of inner sense is therefore only maintained if the relation between active subject and passive self is not settled through a grounding identity for either one (which it is in Kant's account of moral action and of passive receptivity). Once again, Deleuze has placed a productive and genetic paradox where once there were philosophical foundations. This paradox – the destructive and constructive interplay between subject and self – ensures the openness and variability of subjects and selves, but also their necessary failure as foundations.

The criticism of Kant gives Deleuze a set of indicators for answering the question of why the form of time must be pure and empty in the fracturing of the subject through its dependence on the self as another. The self cannot be identified either through an appeal to an external guarantor (hence the necessity of the death of God), nor through an appeal to an enduring form of receptive passivity. Therefore, first, this other must be radically other, that is, resistant to identity and to representation. Second, time itself cannot be determined in such a way as to allow a subsequent determination of things in time. Third, time cannot be identified through a further external reference such as Descartes's and Kant's reliance on God, though in different guises (God as perfect guarantor and God as summum bonum). These indications explain Deleuze's at first sight very strange reference to Hölderlin and to his work on Kant. According to Deleuze, Hölderlin's work on Kant and on Sophocles shows a way to the pure and empty form of time through 'the continuous diversion of the divine, the prolonged fracture of the "I" and the constitutive passion of the self' (DRf, 118). Through the diversion, fracture and passion, Hölderlin dramatically affirms the paradox of inner sense.[3] There must be no possibility of a return to a form of an extra-temporal guarantor of identity. The identity of the 'I' must therefore remain fractured through an ongoing process, and the self, as constituted by passivity, must not have a prior identity restricting this passivity, for instance, through the Kantian definition of receptivity.

BACK TO PLATO

Deleuze follows the critique of Kant for not going far enough towards a pure and empty form of time with a reading of Plato. According to this interpretation, the Platonic theory of knowledge as reminiscence, whereby knowledge must be remembered (partly as a solution to Meno's paradox) rather than acquired through the senses, is a forerunner of Kant's situation of the self in time. Deleuze's focus is interesting since it insists on a double feature of reminiscence, as forgetting and remembering: knowledge is recalled, but therefore it must also have been forgotten. The importance of this work on Plato is very great because Plato is in fact much closer to Deleuze's account of time than Kant. Though the latter shows the way through transcendental deductions and passivity in time, the former precedes Deleuze's accounts of the second synthesis of time, in the Bergsonian notion of the pure past, and of the circular nature of time, in Nietzsche's eternal return. This is significant because it raises an objection to Deleuze's move to a pure and empty form of time as the third synthesis of time. Why is the third synthesis necessary when we already have the pure past in which we can situate the self in time, as the most concentrated tip of the expanding cone of the pure past (as studied in the previous chapter)? The simple answer is that this would determine the self in relation to that pure past, a point that Deleuze shows through a further set of remarks on Plato.

The main features of Deleuze's points about Plato divide into two and allow us to understand the construction of Deleuze's philosophy of time. On the one hand, there are aspects shared by the two philosophies: a circular notion of time, a relation between a phenomenal world and a world of Ideas (the actual and the virtual in Deleuze), and the foundation of time through the past rather than the present (explaining the priority of reminiscence for Plato and the necessity of the second synthesis of time for Deleuze). On the other hand, there are important differences. Plato's circle is organ-

The third synthesis of time

ised through layers of resemblances culminating in the 'in itself' of the ideal. Things resemble the ideal to greater and lesser extent and those most capable of returning are closest to the ideal. This allows the soul that has preserved the in itself or the ideal as such to escape from the circle. It does not return but is always the same ideal other things return to – to greater or lesser extent in relations of resemblance. In Deleuze's circle, taken from Nietzsche's eternal return, there will be no order of resemblance, no ideal or in itself and no escape through true knowledge.

For Plato, life is circular because souls go through a process of metempsychosis or a circle of death and rebirth (as already discussed by Deleuze in relation to the second synthesis and in the previous chapter of this book). But each soul can return to the ideal or immortal part of its former incarnations through knowledge of the Idea, that is, through a return of the Idea as the same origin from which all phenomena decline. This means escaping the circle, because the ideal remains the same and hence does not revolve. So only those things that resemble – simulacra – revolve. Deleuze will return to this important point on simulacra later in *Difference and Repetition*, also in the context of circles but with the added account of Nietzsche's eternal return. Deleuze's overturning of Platonism is thus to put the fractured subject and all its simulacra in a circle with no identified centre – a paradox we shall revisit in the next chapter on eternal return. For Plato, simulacra return as secondary beings in relation to the eternal ideal. This is exactly what Deleuze wants to avoid, since it imposes identity back on to the self and forms a false hierarchical circle based on the return of the same, where a turn of the circle is only indexed through resemblances between lesser or illusory states themselves organised in layers. Plato's circle is not really a revolving process; it is a fixed disk with a hallowed centre and a damned periphery.

The core of Deleuze's argument is more complex than this fairly standard reading though, because it studies Plato's circle in terms of the concepts of foundation and founding that Deleuze appealed to in his arguments for Memory or the second synthesis of time. In orienting the circle according to orders of resemblance, Plato turns the founding Idea into a foundation.[4] This foundation is then not past, in Deleuze's sense of something that makes the present pass through a process of transformation thanks to a capacity to change (Deleuze's pure past). On the contrary, the past as foundation is still a present, because it is a present that has become past. However, it is not a present like any other, since all other presents exist in relation to that foundation. So the present as foundation has to be a mythical present: '[. . .] the Idea is like the foundation from which successive presents are organised into the circle of time such that the pure past that defines it must necessarily still be expressed in terms of the present, as an ancient *mythical* present' (DRf, 119). Deleuze's objection to Plato is then two-fold. First, the past becomes the ground for identity and representation in the circles of resemblance of presents to mythical present. Second, this mythical present subsumes the past itself to representation and identity, and thereby it turns the past back on to presents that it is supposed to found, in the sense of make pass or dissolve in terms of identity: 'This is the insufficiency of the founding, to be relative to what it founds, to borrow the character of what it founds, and to be proven by them' (DRf, 119).

Deleuze does not make explicit the consequences of this work on Kant and Plato for his definition of the third synthesis of time at this point of *Difference and Repetition*, so it is worth drawing them out. The problem of identity and representation of the past is not fully solved by a pure past that eludes representation if this past becomes identified in relation to a new present, either through the identity of the actions of a subject, or through the passivity of a self, or through resemblance of new presents to a former mythical one. The circle of time, the way the past returns, must then itself be thought in such a way that when the past returns with a new present it does so free of any determinations. The new present must therefore be new in a very radical manner, that is, it must be completely undetermined, yet determinable. This explains, though in no way fully at this stage, why Deleuze appeals to a pure and empty form of time: a time that renders the new present determinable yet undetermined. The problem he has set himself is how to relate the new present to the pure past and to the present as synthesis without determining any of them. This is why he adds the work on Plato to the work on Kant: it allows him to introduce the idea that time is circular with the paradox that this circle must not involve the return of identity and representations, of the same and the similar.

THE PURE AND EMPTY FORM OF TIME

Deleuze's work on Descartes, Kant and Plato is essentially critical and still leaves many important questions without full answers. What exactly is the third synthesis of time, the pure and empty form of time? Why does Deleuze's work on this synthesis not situate the subject and the self as foundations for his philosophy? Why is the third synthesis necessary rather than merely desirable or even convenient for Deleuze's model?

Two concepts allow for an answer to these questions. They remain in the background of Deleuze's account but are implied by its content throughout: open determinability as a given and drama as form. As we have already seen, after his work on Kant, Deleuze moves into Hölderlin's work on Sophocles to which he adds a discussion of Hamlet. The significance of this move lies in the following argument. The self is determinable in time through the dramatic actions of the subject. This subject, though, is fractured due to its dependence on the passive self. The subject is then not the foundation for the determinability, but rather the vehicle whose actions are further conditioned by passivity, which Deleuze defines through dark precursors and larval subjects as emerging singularities in the self (as studied in Chapter 2 above, on the first synthesis of time). The relation of dramatic actor to determinable self is hence circular. The actor is conditioned by the self that is created by the actor's dramatic action. However, the nature of this circle is to be radically open; something Deleuze traces later in *Difference and Repetition* through studies of drama, animal evolution and biopsychic evolution revealing the openness of determinability. Though the subject, or in fact any selective action, is determined by the passive self, or any relation of determination through a synthesis of the past, activity is still undetermined in relation to novel differences that can be introduced into ongoing processes. In terms of the core principles of Deleuze's account of the third synthesis of time, this is rendered as the principle that only difference returns and never sameness. This is the synthesis's capacity for genuine novelty, rather than a repetition of the past. The third synthesis of time is the condition for novelty, which is why it must be pure and empty.

The significance of Deleuze's reference to Hamlet and to Oedipus in his philosophy of time is then not only about what they say or do, but how they say or do it. It is not only that for them time is a maddening adventure, whereby any effort to keep it the same, or to bring order back to it, ends in tragedy. It is also that time has to be dramatised, that is, it must be assembled through a novel creative attempt where past, present and future are unified but only also as divided around an encounter with the new. The third synthesis of time is then necessarily determined as a drama, though this necessity is deduced as a pure form. The speculative moment in Deleuze's method in his deduction of the third synthesis of time therefore lies in the encounter of the new, setting time within tragic drama, that is, as a time where the new undoes an effort to live with memory in the living present. The dramatic phrase uttered by Hamlet that Deleuze focuses on is 'the time is out of joint' (*Hamlet*: I, v, 189); in the French translation used by Deleuze, '*le temps est hors de ses gonds*'. '*Gond*' means hinge, as in door hinge, rather than simply joint; so some of the richness of the original is lost in this more precise word. According to Deleuze, in line with a standard interpretation of Hamlet, time has become unhinged or disjointed (in the way the ball and socket of a shoulder might come separated). Hamlet's task will be to reset it.

Yet Deleuze does not follow through to that task and instead develops a line of thought about time out of joint.[5] The Latin root for '*gond*' is '*cardo*' the hinge or axle, the north–south axis in a city, or orienting direction and cardinal points for a circle, for instance north on a compass. The number of revolutions of a circle can then be measured thanks to cardinal points, for instance, in the number of times a clock passes midday or a horoscope passes a birthday. The passing of time, as much for the soul as for the world, he says, is measured thanks to such cardinal points. When these go missing time goes out of joint and becomes maddened and disordered, as it does for Hamlet and the kingdom of Denmark when he receives the news, from his father's ghost, of his uncle's betrayal: 'Thus was I, sleeping, by a brother's hand / Of life, of crown, of queen, at once dispatched' (*Hamlet*: I, v, 74–5). There are events that set time out of joint. But what is this time out of joint? Is it no time at all? Or is it something other than time? Deleuze's answer comes out of a study of what happens when time loses its cardinal points, for instance, when the rightful succession of kings measuring epochs for a kingdom is upended by a betrayal, the return of a ghost seeking revenge and an illegitimate ruler. He claims that such time is empty, that is, empty of the contents given to it by cardinal points.

What are such contents? They are a cardinal numbering (first, second, third) and the significations associated with them (original, closest copy, distant copy, true successor number, false successor number). Time out of joint or the third synthesis of time is therefore empty, in the sense of lacking cardinal numbering, and pure, in the sense of lacking hierarchies associated with numbering. We can understand this disruption further by returning to Deleuze's critique of Plato. In the Platonic context, time out of joint means that the ideal is removed from the circle; we therefore are left with simulacra with no principle of resemblance. For Hamlet, once

The third synthesis of time

his father's ghost has spoken, the numbering and legitimacy of kings is out of joint and his time becomes empty (there is no next numbered ruler) and pure (there is no legitimate ruler). Hamlet's task is then to revenge his father by killing his uncle, thereby re-establishing numbered order and legitimacy based upon it.

So, according to Deleuze's work on Hamlet, an event such as the return of a king as ghost breaks the circle of time by breaking its numbering. Time is then freed of the kinds of content one finds on such circles (dates) and purified of the kinds of values associated with them (origins, copies and ends). This does not mean, though, that time becomes nothing. Instead, it changes from a cardinal time numbering revolutions of the circle, to an ordinal time setting events in order: before and after the appearance of the ghost, in Hamlet's case. This ordering is in fact a cut or caesura, that is, once again, the introduction of an asymmetry in time around what Deleuze calls a formal and static distribution. This is a very technical use of both terms in relation to the other two syntheses of time and hence explains the difference with them. The first synthesis of time is constituted by multiple syntheses associated with living presents. Each present contracts the past and the future and hence each one makes a distribution of past and future events; Deleuze therefore calls this distribution empirical, because it is discovered and made anew with each living present. This is not the case for the third synthesis of time, which is formally cut at whichever present into an ordering of before and after. The second synthesis of time is dynamic, in the sense that it is an ongoing variation of relations in the pure past; the past is always synthesising its relations in different ways and is never static (against usual intuitions about the past, as we saw in the previous chapter). However, the third synthesis is static because its sole characteristic of order always remains the same, the before and after of each cut always remain.

This description of the third synthesis raises an important objection allowing us to understand its deduction and Deleuze's method better. As we have seen, Deleuze's moves towards the third synthesis of time grow out of relatively cursory studies of Hölderlin and Shakespeare. How then can he claim that the third synthesis involves a formal cut, when in fact it is deduced from a somewhat narrow dramatic event (the appearance of a ghost to the Prince of Denmark)? The answer is that the dramatic event is an example communicating a formal and static state of time rather than its ground or foundation. Deleuze is giving us a formal definition of drama in relation to time. The example of Hamlet allows us to understand how any passive situation in time rests on a formal process, a cut or caesura in time, and a static property, an ordering into before and after. This then leads to an answer to one of the first questions this section began with. The third synthesis of time is not based on human experience. It is rather a speculative claim about time based on the disruptive appearance of the new. If novelty is accepted in any process (animal, vegetable, mineral) then a third synthesis of time is implied as condition for the new. This third synthesis is a cut and an ordering. In turn, this means that any interpretation of the fracture of the subject as the fracture of the human subject is mistaken. It should instead be read as implying that any activity is also passivity and fracturing of the subject of that activity, where active means selecting by introducing a novel difference, passivity means determined by syntheses outside the control of activity, subject means process of selection, and fracturing means dissolution of the subject through the syntheses it is passive to.

So it is not any particular drama that makes the third synthesis of time, it is rather that the third synthesis of time is the condition for any particular drama, that is, the conjunction of an effort towards numbered circular time, in which worlds can return, and its fracturing: 'The caesura, along with the before and after that it orders once and for all, constitute the fracture of the I (the caesura is exactly the birth point of the fracture)' (DRf, 120). I have rendered caesura as cut in the above, in order to capture the division brought about by the third synthesis (cut is from the Latin root '*caes*'). However, this misses some of the elegant aspects of Deleuze's choice of word, since the '*césure*' in French is the division of a poetic hexameter, where the verse is split in two unequal parts of five and seven by a pause. He is thus insisting on the asymmetry of the cut and the difference between the before and after.[6] He is also picking up on a remark by Hölderlin such that time stops rhyming due to an unequal distribution.[7] Note that the passage just quoted assigns the fracture not only to the cut but to the before and after (this is lost slightly in the singular tense of 'constitutes' in the current English translation and in its use of 'ordain' instead of 'order'). What this means is that the fracture is not only down to the pause or cut, but also to what it distinguishes. The third synthesis is not only a break, a wound, or division. It is a division into two unequal parts.

A further misunderstanding can now be dissipated. It could seem from the emphasis on the drama of the cut, that the ordering of time and the cut take place in the present and thereby order time, but only each time from the point of view of the present. This

The third synthesis of time

would mean that the third synthesis of time would resemble or even be equivalent to the living present associated with the first synthesis of time, but with the added feature of novelty. This is mistaken because Deleuze's point is that in the third synthesis every event on the line of time has an ordering of before and after in relation to it. Every event is such a cut. It is important here to use the term event, rather than point, because the cut is not point- or instant-like in Deleuze's treatment. Instead, he follows up his remarks about ordering and the caesura with another feature of the third synthesis of time: time is defined as an assembly and as a series. His definition and deduction of these two features follow from a close analysis of the process of the cut and ordering.

It is particularly interesting to study this analysis of the processes of assembly and seriation because it demonstrates the speed, power and scope of Deleuze's close-range analytical and logical thought. It also demonstrates the great contrast between two speculative views of the ordered line of time. On the one hand, this line can be seen as a line of parts ordered according to prior and successor points in time. Each point on the line has a set of points before and after it. This is not Deleuze's model. Instead, the caesura is an event and has a depth to it. It is not instantaneous but rather must be considered with its effect on the points before and after it. This is why the caesura implies a drama: it divides time such that a drama is required to encompass this division. This event-like and dramatic division is in contrast with the thin logical point and set account of the line of time where an arbitrary point is taken on a line and every point before it is defined as before in time and every point after as after in time. In terms of this contrast, it is worth noting that the current translation uses 'totality' for the French '*ensemble*'. This misses the process at work on Deleuze's thought and model. In fact he has analysed the idea of the cut and ordering and tried to explain how they can cut into and order the whole of time, or the totality. So the *ensemble* is rather a process of gathering and tearing apart. Why is this? It is because the whole of time is ordered, but it is ordered differently, that is, into before and after. The events of time are gathered in an ordering, but they are also torn apart, since some are irreparably before and others irreparably after the dividing event. Deleuze's astonishing speed of thought and extraordinary insight come in here. His analysis has taken but a few sentences, but it responds to a devastating problem that haunts the logical before-and-after time line treatment. What is the difference between before and after on this line? How can we make this difference without falling back on to an idea of time organised by the difference between past and future for a human or at least lived present?

The difference between before and after in Deleuze's ordered time is established by a drama. This does not have to be a human drama or even a drama where the active subject is some kind of conscious decision-maker. Instead, Deleuze explains order through the concept of a series. This is not the same concept as the one used in his *Logic of Sense,* but there are connections, notably in the role of expression and dramatisation in that book. A series is ordered by a dramatic event according to the following principles: 'before' is determined in accordance with the assembly of the line of time in an image operating as a symbol for that assembly. When the image is posed as 'too great for me' for a given event, then we have something before. When the image is not too great, then we have the after. Now at first sight this seems to confirm all the worries about the human- and subject-centred properties of Deleuze's account of the third synthesis of time. But that is not the case, once we realise that the image applies to any novel process, for instance, when a virus mutates and achieves something 'too great for it', or when pressures on rocks transform an organic layer caught between them into something new. Deleuze puts quotes round 'too great for it' in order to draw attention to the point that there is only an analogy operating through the phrase around the figure of a drama reflecting on things that are too great:

> In effect, there is always a time where the action in its image is posed as 'too great for me'. That is the *a priori* definition of the past or of the before: it does not matter whether the event is itself accomplished or not, that the action has been done or not; it is not according to this empirical criterion that the past, present and future are distributed.
> (DRf, 120)

The current translation gives '*l'action dans son image est posée*' as 'at which the imagined act is supposed' (DRe, 89). This is too psychological a rendition, since Deleuze never appeals to the faculty of imagination. However, in defence of that translation, could it not be countered that it is plausible because it is consistent with Deleuze's reliance on human drama and on images? There are a number of ways of responding to this point and it is important to do so because the question demands a wider response to the centrality of human drama in Deleuze's account of the third synthesis.

First, the focus on human imagination and particular instances of drama fits very uneasily with Deleuze's points about an a priori

The third synthesis of time

definition. If the past or before was determined each time in the imagination, then we would be dealing with an empirical psychological test; something that is explicitly rejected not only in the above passage but also in the contrast with the function assigned to the empirical in the first synthesis of time. Second, psychological explanation is avoided at every turn in *Difference and Repetition* because it fails to grasp the importance of habit and extra-mental processes (as we saw in the study of the first synthesis of time). Third, Deleuze uses the term image in a technical manner, partly indebted to Bergson. For Deleuze, the notion of image is not one of a mental image, but rather one of a reductive yet necessary process of assembly (a point that comes out very strongly in chapter III of Deleuze's book, 'The image of thought'). In that chapter, as elsewhere in his work, the image is not an image for consciousness and thought is not conscious thought in a mind. However, the difficulty remains that Deleuze is referring to actions by subjects and to what certainly appears to be a conscious awareness of an image that is 'too great for me'.

The solution to this difficulty lies in a closer analysis of that central phrase in relation to the puzzling claim about the a priori nature of the sealing of before and after. When we first read the phrase, we read it as a first-person statement – perhaps one in an inner dialogue, *it is too great for me* . . . Yet, as we have seen, this cannot be consistent with Deleuze's other statements. Instead, the phrase should be read according to the following paraphrase: 'the action in its image is posed as too great for the self'. Deleuze's point is therefore, once again, a formal one. There is a necessary assembly of time implied by any possible cut in time. This assembly depends upon an image standing as a symbol of the times assembled. This image is necessary for action in relation to a novel event, but the image is too great for the self, that is, for the underlying processes the subject is passive to. It exceeds those processes as past or before. None of this depends upon an empirical experience. It is rather a formal connection between processes of cutting, assembly and setting into series in relation to the future as novelty:

> As for the third time that uncovers the future: it signifies that the event and the action have a secret coherence that excludes the coherence of the self. It has become equal to them and they shatter it into thousands of pieces, as if the bearer of the new world were taken up and dissipated by the shattering of what it, in the multiple, gives birth to. The self is equal to the unequal in itself.
>
> (DRf, 121)

The third synthesis of time leads us to the future. This future, though, is not a dimension of the present or of the past. Replicating the move from the present to the past in the second synthesis of time, where the former became a dimension of the latter, in the third synthesis the past becomes a dimension produced by the future. This production takes place in the cut, assembly and seriation of time. In the cut, the subject is undermined by the self, but then the subject returns as the actor reuniting past and future in an action that symbolises them. This unification though also makes the past and the future by shattering the self, the vehicle of the past in the present action. So the multiplicity of the self, the many selves underlying the subject of an action, is itself broken and remade in the third synthesis of time.

We can now return to the three questions posed at the beginning of this section. The third synthesis of time, or pure and empty time, is a cut, an assembly, an ordering and a seriation. It is deduced as an a priori condition for action, which in simple terms claims that any novel action depends on a cut in time. This cut though must also assemble what comes either side of it. This assembly is itself dependent on a putting of time into an order of before and after the cut. The third synthesis of time is therefore a division of time and an ordering of time. This ordering though is also a seriation; it distinguishes the before and the after, rendering time asymmetrical. This complex third synthesis is the time of the future making the present and the past, which become dimensions of it, because the action it is posited upon is essentially determined by an open future. Neither the subject nor the self is a foundation for Deleuze's philosophy, because in the third synthesis of time they are both ungrounded. The subject is divided by selves it is passive to. These selves though are shattered by the structure of the third synthesis that makes them past in relation to a future they cannot determine. All of these components of time are formal and deduced as necessary conditions for any novel action. So though Deleuze depends upon examples and vocabulary from drama to introduce and explain the third synthesis of time, this drama is strictly defined formally as the cut, assembly, ordering and seriation which can be explained by reference to Shakespearean drama but is not derived from it. As Deleuze insists all the way through his work on the third synthesis of time, it is a priori and not dependent on empirical observation.

The third synthesis of time

HISTORY, REPETITION AND THE SYMBOLIC IMAGE

Once Deleuze has defined the third synthesis of time, he proceeds to a reflection on the implications of his new philosophy of time for repetition in relation to history. This is important for at least three reasons. First, it allows him to respond to another objection to his deductions, on this occasion around the concepts of image and symbol used to account for the assembly of time in the third synthesis. Why are these concepts not human oriented and subjective? Why do they not make this new philosophy of time dependent on historical accounts of images and symbols? Second, it allows him to connect this philosophy to his account of events in its important roles in *Difference and Repetition* and *The Logic of Sense*. Third, the reflection on history is the place where Deleuze connects the third synthesis of time back to the overriding concept of repetition, a connection made for each of the syntheses. The third synthesis of time is, like the other two, a form of repetition. The question is 'How and with what repercussions and dependencies in relation to history?'

Two familiar theses about history are not what Deleuze means by repetition in the third synthesis of time or in history. He is not claiming that history repeats itself and he is not claiming that we, or any actors, are destined to repeat history. His view is exactly the opposite and it is important to understand why. The root of the answer lies in the a priori nature of the third synthesis and in its relation to repetition in the first and second syntheses. The past is repetition but only as pure past, that is, it is only repetition of dynamic changes of relations in the past, rather than repetition of the same events. We repeat the past by changing it. The present is repetition, as synthesis through habit, but this synthesis is a transformation, a metamorphosis, of earlier series. The future or the third synthesis of time rests on both these other syntheses. They are the conditions for it, though in very particular ways that will be studied in detail in the next section. There is no past to repeat in the future, since the past makes the present pass in such a way as to make its return impossible. There is no possibility of repeating in the present as directed towards the future, because any action in the present must transform the series it contracts. However, the past and the present, as rendered in the first and second syntheses, also rest on the future and on the third synthesis, because it is the condition for the new in both. If you take a case from the past and attempt to repeat it, you will only be repeating a false representation. When you make an effort to repeat something from the past, that very effort is dependent upon changing the series of past events that condition it. So instead of guaranteeing a repetition of past events, the past and the present are in fact conditions for the necessity of novelty. Given these impossibilities and these conditions, how can we speak of repetition at all in relation to the third synthesis? What continuity does Deleuze allow for history?

The answer to these questions is indicated in italics in the paragraph where Deleuze explains the connection between the symbolic image operating in the third synthesis of time and repetition: '*Repetition is a condition for action before it is a concept of reflection*' (DRf, 121). This statement rests on all three syntheses. First, action rests on repetition in the present because any present act is a passive synthesis that contracts past and future events in the present through a metamorphosis. As we saw in the chapter on the first synthesis of time, here, any present act rests on habit, which is itself a form of repetition (as demonstrated in Deleuze's studies of repetitions in Hume and Bergson). Second, the past as dynamic repetition of past relations is a condition for the present, as active and passive. Third, the new as produced in a present act and conditioned by the third synthesis of time as cut, assembly order and series is itself dependent on repetition as the eternal return of difference. Thus any reflection on a repetition presupposes a prior and deeper form of repetition, or in fact, three prior interrelated forms of repetition. The difference between repetition for reflection and repetition as condition for any action is crucial to Deleuze's argument.

When the historian makes the claim that an event has been repeated or that history generally repeats itself, the claim is about repetition of same events. Yet, Deleuze has shown the claim, the events and even sameness in general to be dependent on forms of repetition as variation: variation in synthesised series, dynamic variation in the pure past and return of pure difference as condition for the new. This explains Deleuze's statement on the analogical form of the historian's claim: 'It occurs that historians search for empirical correspondences between the present and the past; but however rich that is, this network of historical correspondences only forms a repetition through similarity or analogy' (DRf, 121). The argument here is not about historical facts. It is a transcendental argument about necessary conditions. Any claim about empirical repetition of the same must be analogical because there are in reality no events with an identity capable of sustaining a claim to sameness. This is because real events have repetition as variation or differential repetition as their conditions: 'There are no facts of repetition in history, but repetition is the historical condition under which something new is effectively produced' (DRf, 121). In turn, this argument explains and also clarifies Deleuze's statement that the actors of history are determined to repeat before any historical reflection on repetition. Any act is determined by repetition through the three syntheses of time. Once again, any translation into English is full of risks, since a translation is in danger of putting the onus on a conscious decision to repeat and to repeat an empirical situation due to the ambiguity of the verb 'determined' and the apparent clear meaning of 'for themselves': '[. . .] it is in the first place for themselves that the revolutionaries are determined to lead their lives as "resuscitated Romans" [. . .]' (DRe, 90). There is no necessity for awareness of repetition here, nor any self-attribution of repetition of a historical event. The actors have not embarked on a course they are committed to accomplish. It is rather, much more radically, that any historical actor must repeat the past and past lives as figures from the past. The actor need not represent such figures, in fact cannot truthfully represent them, but instead repeats their past conditions as conditions for the new. We are truly all the names in history, before we represent any chosen name to ourselves. Any historical actor, though, only repeats the past by creating the new.

Deleuze's discussion of history in the chapter on repetition for itself then passes to one of the first accounts of the role of eternal return in *Difference and Repetition*. I will return to this discussion in the next chapter on eternal return. My concern here will be to follow through a more enigmatic and less well developed side of Deleuze's thought in relation to the third synthesis of time: the symbolic image associated with the assembly of time. The image appears in two ways in the work on the third synthesis, as concept and as a particular set of images. These images are highly enigmatic for three reasons. First, they do not appear to fit well with Deleuze's methods of close logical and conceptual examination, transcendental deduction, critique based on transcendental arguments and historical situation, both philosophical and properly historical. Second, as was noted in the questions at the outset of this section and throughout this chapter, the reference to images seems to draw Deleuze into the subject as interpreter and producer of images and therefore into the problem of subjective accounts, and of narrative and hermeneutics.[8] The image is the moment where Deleuze's formal account of the third synthesis as drama seems most at risk of falling back on to an empirical and subjective account of time as human drama. Third, the images themselves are unusual and taken from often rather obscure contexts with little scene setting or explanation. Nonetheless, I will argue that images are an important part of Deleuze's work, exactly because they allow for responses to these problems.

The symbolic image enables processes of assembly, ordering and seriation. It is necessary because any cut in time presupposed by the new is a bringing together, an ordering and a setting into irreversible series of the whole of time. The difference between the order and series lies in this irreversibility or asymmetry of time. For Deleuze, time is not only ordered into before and after and around any event, it is also set into a series such that there are intrinsic differences between the before and the after. All these processes need to be determined in an action in relation to the new, where this action is a selection. This is not necessarily a conscious choice, since the sufficient condition for action is a passage from a set of determinations to another one, thanks to the return of pure difference. Any given cut in time must be assembled in relation to action through an image because the action is neither a synthesis of the whole of time, nor a satisfactory representation of the whole of time. The subject of the action is passive in numerous ways through the passive syntheses of the first and second syntheses and through the fracture in the subject implied by the third synthesis. This means that the assembly can only be symbolic. Yet this raises a very severe critical point. Does this make representation a necessary and central component of the third synthesis? The answer is no; that is, Deleuze uses symbol and representation in very precise ways in chapter II of *Difference and Repetition* and the terms cannot be seen as equivalent. Representation imposes four rationalisations on different forms of determination according to Deleuze's reading: identity, analogy, opposition and resemblance (DRf, 44–5). Representation demands identity in the representation and the represented thing. The differences between representations and between things are then thought in terms of oppositions. The relation between thing and representation is analogical and must be judged in terms of resemblance. This is not how the symbolic image works in the third synthesis, as can be shown through a close study of the key definitions of the symbolic image and through a study of Deleuze's examples.

I will study the key passage in detail, in order to draw out the full reasoning behind Deleuze's argument for the symbol and its resistance to charges concerning representation and subjectivity. More broadly, I will consider questions of interpretation such as whether the place of the symbol in Deleuze's work gives a central role to meaning and interpretation in his philosophy of time. Here is the full passage:

> Firstly, the idea of an assembly of time corresponds to this: that any caesura whatever must be determined in the image of an action, of a unique and formidable event, adequate to time in its entirety. That image itself exists in torn form, in two unequal parts; nonetheless it draws together the assembly of time. It must be called symbol, in function of the unequal parts that it subsumes or draws together, but as unequal to one another. Such a symbol, adequate to the whole of time, is expressed in many ways: to make time out of joint, to make the sun shatter, to throw oneself in the volcano, to kill God or the father.
> (DRf, 120)

It is perhaps best first to register the main translation issues. As I have already noted, I prefer 'assembly' to 'totality' since the latter term elides the process and does not fit the later remarks on inequality, adequateness and expression. The assembly is a transformation that is neither equivalent to a totality, nor a totality in itself. The current translation gives '*césure quelconque*' as 'caesura, of whatever kind' (DRe, 88). According to my interpretation, this in not quite right because Deleuze's point is not about kinds of caesura, but rather that *any* caesura in time is determined in a symbolic image. Time, in its third synthesis, is nothing but caesuras into before and after. However, this is an extraordinarily strong claim when we match it to the idea of the symbolic image. It invites disbelief due to the mismatch between the any and the definition of the symbol as 'formidable event'. On the one hand, it is more consistent to assume that any caesura whatever must be determined by an image, since otherwise time would consist of stretches without cuts; it would be intermittent time dependent on the symbolic reception of great events. On the other hand, however, this also means that there will be caesuras which appear to be unsatisfactory in terms of their suitability to symbolic images and formidable events. When the ghost first appears with his terrible message in *Hamlet* it is fairly easy to conceive of this as a symbolic break around a symbolic image: the ghost carries forth the past into a present that it tears apart. But what about the ghost's repeated groans offstage during the same scene? They certainly move the action on and indeed are designed to be prompts to action, to a swearing to revenge. Yet is each one accompanied by a symbolic image assembling and dividing time?

The answer to these questions and conundrum lies in the

necessity of the image and what it is necessary for. The image of an action, of a unique and formidable event (the English translation reverses action and event in the clauses) must be adequate to time in its entirety (given as a 'whole' in the translation). Every event in time must be situated in relation to the caesura as before and after. Deleuze cannot appeal to a formal linear model of time for this, for instance, by representing all events as points or stretches on a line and then by making the claim that, once the line is divided at a point, everything on the line that precedes it is before and everything after it is after. This is because the caesura itself would not determine the before and after. These would depend on the formal representation and there would be no difference between any given cut and any other, and no connection to an action, a synthesis, an assembly of any singular kind. Time as formal representation would take over from time as multiple processes. Yet Deleuze has many critical points to make against such representative views of time, for instance, in the difficulty that though we call different segments of the line before and after, there is no other formal difference between them and hence no asymmetry. Deleuze sets out such asymmetry as a formal condition for the new and for action. Whether he is right to do this is important for the argument and this question will be returned to and has already been considered in relation to the speculative nature of Deleuze's method.

In contrast to representative models of time, when Deleuze says that the image must be adequate to time in its entirety, he neither means that it is the same as it, nor that it corresponds to it perfectly, nor that it is an analogy for it. Instead, the term 'adequate' is used by Deleuze in a technical sense that can be traced to the work on Spinoza he was doing concurrently to *Difference and Repetition*. Long sections from *Expressionism in Philosophy: Spinoza* are dedicated to the concepts of adequate and inadequate. It is always made clear there that 'adequate' in Spinoza must be understood through the concept of expression rather than through a notion of equivalence or of correspondence. This reading is confirmed by the use, in the passage from *Difference and Repetition*, of 'expression' in relation to the adequate symbolic image: 'Such a symbol, adequate to the whole of time is expressed in many ways' (DRf, 120). In the Spinoza book, Deleuze explains that 'adequate' means 'the internal conformity of the idea with something it expresses' (Deleuze, 1990: 118). However, in Spinoza, this something is the idea's cause, something that it cannot be in the case of the symbolic image, since the whole of time is not the cause of the image, a term never used in these sections of *Difference and Repetition*. In its place, we have the concept of determination and a specific form of determination as a caesura and processes dividing, assembling and ordering time in its entirety. Thus the symbolic image must express these processes in relation to the whole of time in its internal conformity. This is why it must be formidable and a great event, not through any sense of scale or horror, but rather in its range, internal division and internal asymmetry.

We can now see that the symbolic image does not imply an external sense of why it is formidable and great. Instead, it will be a matter of ensuring an internal conformity between an action (a selection and the return of pure difference) and the expression of the entirety of time in relation to that action. Any caesura implies the potential for such an expression. However, it cannot imply a perfect expression or one with a final identity. This is because the expression must also be adequate to a fracture and an inequality: the fracture of the subject and the inequality of before and after, as defined by the phrase 'too great for me' studied earlier. This also explains why Deleuze insists that the expression is through a symbol, that is, in terms of its ancient Greek roots, to a sign made of two broken parts (for instance, the symbolon as divided coin that can be reunited as a sign of trustworthiness, or the division of a complete being into male and female parts that then search to be reunited in love, in the Platonic myth of the symbolon from the *Symposium*). The symbolic image must itself be broken or fractured, despite its power to unify. It cannot be a representation because it lacks each of its characteristics (identity, analogy, opposition and resemblance). The symbol has no final identity because it is composed of two unequal parts that cannot be subsumed into a whole resolving the inequality. The symbol does not operate by analogy, but rather expresses the entirety of time – and must do so in many different ways in accordance with each caesura. Such symbols are not to be analysed in terms of oppositions, since they do not represent the same times according to similar identities, but rather each renders different times through different and necessarily incomplete acts. A symbol is not judged according to criteria based on resemblance, but rather on the singular tests of how it draws conformity between the act and the times it subsumes without either identifying them or erasing their differences. So the adequacy of the symbol lies in how well it draws together the entirety of time while still maintaining the inequality of before and after, and the fracture of the subjects creating the new within the time it assembles.

It is only once we have understood the role of the symbolic image and the demand for adequacy in relation to the cut, assembly and series of time that we can resolve an apparent anomaly in Deleuze's examples for such symbols. This is because the most adequate symbol, on its own terms, is associated neither with Oedipus, nor with Hamlet, nor with Empedocles, but rather with a nameless being, a man without qualities, a plebeian character or 'already-Overman whose scattered members gravitate around the sublime image' (DRe, 90). This theme of a lack of identity and qualities, of a becoming imperceptible, will take on greater significance after *Difference and Repetition*, for instance in *A Thousand Plateaus*, but it is already implied in the earlier book and accompanies Deleuze's critique of romantic irony. To truly be a symbol we must erase any sense of the return of an identity, for instance, in the romantic image of the subject striving hopelessly against the world. We must also erase any sense of a representative subject, the extraordinary hero standing for all of us through exceptional qualities.

Deleuze moves through three stages in his work on the symbolic image. Oedipus and Hamlet are adequate to the first two stages as they are thrown into and torn asunder by the past, then drawn into the present where their acts testify to the cut in time, to the potential for a break with the past. However, in relation to the future, where only difference returns, neither character is sufficient, since both are tragic in drawing all the action and all of time on to their activity and identity, in tension with the necessity of a fractured subject and exploding self. Instead, the most adequate symbol will be a figure with no self and no identity, with a 'secret coherence' (DRf, 121), meaning a coherence in faceless multiplicity rather than identity. Deleuze will often return to this most adequate form of symbol, for instance, in his reflections on Bartleby in *Essays Critical and Clinical*, or on Riderhood in 'Immanence: a life . . .'. Note, though, that this has to be the most adequate form in each singular case; it is not the basis for comparisons between cases. Imperceptibility takes on different forms and each is only adequate in its own right and not in relation to a type.

PAST AND PRESENT AS DIMENSIONS OF THE FUTURE

Once Deleuze has set out the main deductions of his philosophy of time, he summarises its three syntheses in a single paragraph (DRf, 125; DRe, 93–4). The point of the summary is to insist upon the main turn of the work on the third synthesis. Just as the move from the first synthesis to the second involved a change in dimensions, where the present became a dimension of the past, in the third synthesis, present and past become dimensions of the future. As such dimensions, the past becomes a condition of the future and the present becomes an agent. What does this mean and what is stake in this shift? Deleuze explains it in terms of the central ideas of foundation and founding. In the first synthesis, the living present is the foundation of past and future, they depend on how they are contracted in a living present and the present is prior to both of them, where prior means that they follow from a synthesis in the living present. *There is no past or future independent of each synthesis in a living present.* In the second synthesis, present and future are founded by the past. Relations between presents must accord with the dynamic nature of the pure past. *There is no present that is not accompanied by a dynamic change in relations in the past that it has no control over. Each present is only founded as passing.* The relation between present and future is founded on the impossibility of a return of any present. *The future is left undetermined by the passing away of each present and the pure past founds the future as open.*

It seems odd, therefore, to think that the past becomes a dimension of the future by becoming a condition. Is not the role of condition the sign of a prior position in Deleuze's transcendental philosophy? This puzzle can be resolved when we realise that what matters is the change when a process becomes a dimension of another. For the past to become a condition and a dimension of the future – when the past was a founding in the second synthesis – is a profound change in priority. This is because when the past is a condition for the future, it is the future that sets what returns from the past. This explains why Deleuze says that the past becomes a condition 'by default'. The past is a condition that fails in relation to the future. Deleuze sums this up in the following way: 'The repetition of the future is the royal one as it subordinates the other two and strips them of their autonomy' (DRf, 125). It is very important to study this statement in detail. It does not mean that the third repetition is independent of the other two as process; on the contrary, they remain necessary dimensions of the future as actor and condition. It does not mean that the other two are completely subordinated to the third; they are only such when they are dimensions of the future, not when it is a dimension of the present or of the past. What it means, then, is that when the present and the past become dimensions of the future they no longer operate as foundation or as founding, that is, the present no longer selects a singular series through a contraction of a singular past and singular future, and the past no longer founds the present and the future through its dynamic relational variations. Instead, both the present and the future become subject to the processes of the third synthesis: '[. . .] the third ensures the order, assembly, series and final goal of time' (DRf, 125).

Thus, in the order of time as set by the third synthesis, any given living present becomes any present whatever, rather than the foundation of time. It takes its place among all other presents in that order. It is assembled with all others, not in accordance with its singular contractions of the past and of the future, but in a symbolic image of the entirety of time and in relation to an action. It is set in a series, that is, severed from its past contractions and set into an asymmetry that is no longer the difference between past and future syntheses for the singular living present but rather a determination through the new and for the new. This means that the goal of time is no longer set in accordance with each living present, but for the new and by the new. The past loses its founding role in each of the processes of the future. When set in order the dynamic processes of the past are no longer determinants of a given present as their most contracted state, but rather have to be taken all together as the entirety of an ordered time. In the assembly and seriation of time, the pure past is subjected to the requirement of the symbolic image, to the point where from the perspective of the future, the pure past itself is such an image. As determined by a seriation, the past is no longer the founding of time through dynamic variations but is replaced by the new, by the future as open. The goal of time is no longer to make presents pass, but rather to make the past pass.

All of this can seem difficult, but it is possible to explain it in three ways. First, we can reflect further on the two concepts that Deleuze uses to express the shift in dimensions: the present as agent or actor, and the past as condition by default. Second, we can think of it in terms of a practical example. Third, we can analyse the few sentences following Deleuze's core statement on the status of the future. I will do all three. When the present passes from the living present to the present as agent as dimension of the future, it could seem that nothing much changes. This is because the subject in the living present is in fact passive, constituted and fractured by passive syntheses of the past. However, the import of Deleuze's remark is not strictly that the subject is an actor, but rather that in the third synthesis, in the future, the actor is 'destined to be erased' (DRf, 125). The processes of the future, like those of the past and the present, turn subjects into actors, but unlike them they do so by breaking even with passive syntheses. Had Deleuze stuck with the first synthesis, his philosophy would have led to a weak determinism constructed around passive synthesis of the past in relation to selections in the living present. The seriation of the third synthesis erases the hold of the passive syntheses of the first. This break is replicated in relation to the second synthesis as condition, because the principle that each passing present is made to pass by the pure past is superseded by that which is wholly undetermined in the present in relation to the future. So, as a dimension of the future, the past is a condition operating by default, that is, it is failed condition, or one that is missing. The future is freed even from the principle that follows from the synthesis of the pure past such that it must pass into memory. The condition operates, but only by default in having no determination on the future.

'No walk has ever been endogenous' (DRf, 132). This remark by Deleuze, shortly after his work on the three syntheses of time, is made in the context of a psychoanalytical and biological study of how children learn to walk (where the emphasis is firmly on the psychoanalytical). However, the claim does not follow from psychoanalysis, but rather from the deductions made in the philosophy of time. Any walk is habitual. It is a synthesis of past events and future possibilities in a living present, which is itself both active selection and passive conditioning. This is the first synthesis of time and a first reason why a walk could never be endogenous. Every walk is in touch with all of the past, so any limit on the processes leading to a walk, for instance, on the limits of the body, is an arbitrary definition of a boundary for a determination of what counts as endogenous. This is because every walk passes into the past, changing in its intense relations to all past events as it slips away. No present act can control this passing, as if history dragged each present into itself by throwing its values and meaning into disarray. This passing away and hold of all the past on any present act is the second synthesis of time. It too makes an endogenous account of a walk nonsensical, since any walk is in touch with and sundered by the whole of the past. This does not mean that it is illegitimate to draw such boundaries. Such selections have to be made and limits drawn for setting out an explanation. The error comes in when a further claim to legitimacy is made for the selection on the grounds that, for instance, a walk is endogenous. This is because then a necessary but arbitrary selection is then justified against the processes going beyond its boundaries; to the point where these processes can be excluded.

Every pace taken by every animal is new. Every roll of every stone is a break with the past. But if no walk is endogenous, because every walk is habitual and in the grip of the past, then does this mean that no walk is truly new? Moreover, if something radically new can occur, does that not support the idea of endogenous development, if the new occurs within the boundaries of a given system and in relation to them? These critical questions account for the care Deleuze takes in describing how the present and the past become dimensions of the future. His answer is that every event is new for two connected reasons: the present is an agent erased by the return of pure difference in any event; the past is a condition of the future only once pure difference has returned. So any habitual gesture and the passing of that gesture are also and necessarily new and driven by the return of pure difference. The new is then not endogenous; it works on and through external syntheses. The past never conditions the future. Every pace taken is new, but it is new for all of the past and for every series synthesised in the present. *However hard the struggles to break with or repeat the past in the present, both fail the test of the new, or the eternal return of pure difference.*

In the remainder of the paragraph summarising his philosophy of time, Deleuze introduces a modernist philosophical programme built around the role of the future in relation to the present and to the past. This programme depends on inverting the sense of repetition from a repetition of the past to one of the future. To repeat the future is not a contradiction. It means to repeat the processes of the future, or to reaffirm them a second time in our acts, such that repetition becomes 'the category of the future' (DRf, 125). According to this programme, the repetition of the present (the synthesis of past events in present acts both active and passive) and the repetition of the past (the synthesis of the pure past) must become stages on the way to a pure difference drawn from them. To do this, there must be no return to a repetition of the same or of the similar and there must be a refusal of cycles that remain too simple (customary motions, memorials and/or claims to be beyond memory). Deleuze is not confusing a time within our volition and a necessary time, here. Instead, the programme is about how best to act in relation to a time that is necessarily a time of the future.

TRANSCENDENTAL DOGMATISM?

Once time has been defined by Deleuze in terms of repetition and difference a novel set of problems opens up for interpreters. I want

The third synthesis of time

to focus on three from very different areas of concern. First, if time becomes a question of repetition, where do we draw the line as far as what constitutes Deleuze's philosophy of time proper? Should all his remarks on repetition – and indeed on difference – be taken as parts of his philosophy of time? How do we distinguish remarks that are proper to time and others that are merely incidental, perhaps just applications or examples? These questions are significant because they draw out a wider problem of interpretation when it approaches thinkers whose works roam widely across subjects and topics. Is there a core to that philosophy? What should we make of interpretations that give greater priority to a connection between the thought and a specific contemporary area – science or politics, say?

This last question leads into the second area of concern. In order to avoid accusations of dogmatism based on a contingent and therefore illegitimate selection of given components for Deleuze's philosophy, I have described it as speculative.[9] Its method for developing its structure is transcendental, though it has empirical moments in its focus on this or that particular topic (for instance, in the first synthesis of time) and in its experimental aspects (for instance, when it tests a new line of thought deducing a set of transcendental conditions for a particular event). However, at its core it is speculative in having to posit certain principles or events without further grounding argument. These speculative assumptions are then tested, for instance, in terms of how they contribute to a consistent theory, as shown in the previous chapter on the first synthesis of time. With the third synthesis of time, the speculative moment lies in assumptions about the new, for instance, in relation to a caesura in time. Although, there might be a counter to this claim, based on a reading of Deleuze's version of eternal return; this will be considered in the next chapter. The worry at this point is not about these speculative moments, but rather about the relation between Deleuze's reading of specific cases of repetition, drawn from biological and psychoanalytical repetition, and his philosophy of time. Does that relation anchor him to a particular scientific moment and indeed to a rather narrow and fragile one, such that his philosophy will be seen as dogmatic once these sciences are replaced by others (assuming that they are, of course)?

This second concern could be viewed within the constraint of a 500-year test. Philosophy is read hundreds of years after it is written, not for strictly historical reasons, but because it is taken to be valid, or true, or of interest in its own right in new epochs. If a philosophy is tied too tightly to an outdated science or to what is considered unscientific opinion, then it will either have to be reread in such a way that the role of the science or opinion is changed, or the philosophy will be devalued in a more fundamental manner: it will become historical, a relic, rather than properly modern and operative in its own right. This is a particularly sharp worry in the case of Deleuze's work in *Difference and Repetition,* because of its mixture of science, art and philosophy with boundaries that are often loosely drawn and with relations of dependency that often seem to put contingent artistic and scientific examples at the heart of philosophical arguments (a risk that we noted in this chapter when reflecting on the role of drama and symbolic images in the third synthesis of time). In the following study of the remarks on repetition, I want to counter such a view and show how Deleuze's philosophy of time can at least offer a reasonable chance of a robust response to a mooted 500-year test. The central claim will be that his work on time is applied to biopsychic examples but does not rest on them. They allow him to illustrate how the three syntheses of time can be applied, but that application is not an important part of the arguments for their truth or validity. Though of course, like any philosophy, those arguments and validity are open to be studied in relation to the sciences in a critical manner. The critique though will be of the philosophical structure itself and of its applicability and consistency with the science, rather than of any dogmatic adoption of a scientific moment.

The third area of concern here is more tightly focused on the analysis of Deleuze's syntheses of time. We saw earlier that the figure of Eros played an important role in explaining how attraction connects the third synthesis of time to the first; that is, put simply, how a living present is also a selection of the past driven by an attraction towards the new in the future. However, in the work on the third synthesis as cut, assembly and seriation, Eros disappears from Deleuze's account, despite having been clearly invoked in the sections prior to the deductions of those processes. This is significant because it raises a contradiction or a paradox in Deleuze's work on the third synthesis. The paradox can be understood best through questions about pure chance and contingency (rather than about chance understood as probabilities).

In separating the future from any final determination by the living present or by the pure past, and even more so, from any determination by representations of actual events that are now past, Deleuze seems to render the future completely unconditioned.

The third synthesis of time

This lack of conditions is what I understand by the openness of the future in my interpretation of his philosophy of time. However, this seems to imply two related difficulties connected to explanation. If the future is completely open, how can we use concepts of selection in relation to later connections made between past, present and future? The future appears never to be selection but always pure chance. Similarly, if the future is completely open and undetermined, how do we explain how an occurring event takes its place in relation to the past and the present according to processes of assembly and seriation? Each completely new event seems to imply a complete jumbling of past events. But since any future event is new in that way, time implies a succession of chaotic reordering, to the point where no order survives the constant arrival of new and disturbing events. In short, how can Deleuze maintain a concept of selection and a concept of continuity, implied by his idea of the role of Eros, with his claims for the new and the eternal return of pure difference?

The answers to the opening two questions from this section can be tracked through Deleuze's critical reading of the role of repetition in Freud, following the discussion of the third synthesis of time. The key point to insist upon is that the reading is critical and that this critique is based on the work on the three syntheses of time. Deleuze follows Freud's work on the pleasure principle not in order to deduce these syntheses, but rather to show how they allow for an explanation of passive syntheses in Freud by mapping them, first, on to habit, then on to pure memory and finally on to Eros. The critical side of Deleuze's philosophy is shown there in terms of relations of necessary priority between processes, not analysed empirically but rather in terms of necessary conditions for given hypotheses. For instance, habit cannot be seen as a consequence of a search for pleasure or displeasure because these already presuppose habitual syntheses. The argument is thus that in order for there to be an idea of an acquired pleasure or of a pleasure to be acquired there must first have been many syntheses connecting many events over time. This passive connection is what Deleuze has called habit in his work on the first synthesis of time: 'On the contrary, habit, as passive synthesis of binding, precedes the pleasure principle and makes it possible' (DRf, 129; DRe, 97). A similar argument is made in relation to a critique of activity as independent of passivity, in terms of the positing of real objects associated with pleasure and desire. Any active positing of the object presupposes passive synthe-ses even when there is a transfer from prior passive syntheses into an activity, such as the search for pleasure in relation to a particular object. Here, Deleuze's work on the pure past supports the critique. The Freudian reality principle, whereby the active positing of real objects, and more generally of reality, sets limits for any pleasure principle, itself presupposing the continued work of the passive synthesis even after this latter has laid the ground for the reality principle: 'Moreover, it seems that the active principle could never be constructed on the passive synthesis unless that synthesis simultaneously persisted, developed on its own account, and found a new formula, both dissymmetrical and complimentary to it' (DRf, 131; DRe, 99). Deleuze's work on time here does not depend on Freud, Klein or Maldiney's work on children's development. It draws support from a critique of it, as it does from a sympathetic study of Lacan's work on real and virtual objects, a few pages further on in *Difference and Repetition*.

This critical work adds nothing substantial to the form or content of the earlier work on the three syntheses of time, though it adds a lot in terms of exemplification and confirmation. So in terms of questions about the extension of the philosophy of time into repetition, we can conclude that it is valuable for our understanding to follow how the philosophy of time maps on to specific cases of repetition, but it is not essential to the arguments for it or to its consistency or definitions of concepts. Therefore, Deleuze's work is not in danger of accusations of dogmatism on the grounds that it rests on scientific views that are prone to be viewed as false or incomplete over time. Neither is it open to the accusation that it rests on patently false science (via Freud, for instance). It is not even open to the judgement that it rests on Freudian or Lacanian theory or philosophy, since again the philosophy of time is developed before any treatment of those theories. Yet there is one glaring possible exception to these conclusions in the work in relation to Eros and real and virtual objects. Is this work dependent on claims that render Deleuze open to accusations of a dependency on empirical evidence or psychoanalytical theories independent of his studies in the philosophy of time?

The key point where Deleuze's account of Eros meets his philosophy of time is where he returns to Bergson and to the pure past within his study of Freud, Klein and Lacan. Eros allows for a return of the pure past in present attractions and selections through virtual objects drawn from the pure past. Any novel drive is not to be explained by a simple attraction to real objects because an active synthesis in the present never presents us with the full object explaining that attraction. Underlying any novel activity towards an object, there are passive syntheses to account for according to two virtual objects, the virtual object as determined by the passive syntheses in the living present and the virtual object as determined by the passing of the present into the pure past. Neither of these can actually be recovered and attraction towards the new is always unfulfilled, not only in a negative sense of seeking something specific that one can never have, but also in a positive sense of being drawn towards something that is necessarily more than can be recognised or grasped. This need not only be viewed in terms of human sexuality. It is rather that any novel drive and binding is made through virtual objects. These objects cannot be real, in the sense of accounted for fully from within a set of known objects, yet neither are they completely unknown or unmanageable.

The virtual object is then a way of solving the paradox of the new presented earlier in this section. It is not fully determined by the past, yet it is also drawn from the pure past. Any attraction and binding to the new is not a leap into complete indetermination and a chaotic set of relations, but rather a partial move through the virtual object. This will allow Deleuze to connect Eros and the virtual object to pure difference and to Nietzsche's eternal return. One side of the virtual object, its novelty, is drawn from the pure past but another, its real part, is drawn from the present. That is why Deleuze describes the virtual object as having an eternal positive lack at its heart. It is neither fully any real object, nor any past object, nor fully pure difference or a set of varying relations in the pure past. It is all of these in an unstable mix, where unstable is also not a negative term, but rather an explanation of the attraction and eternal novelty allowed for by virtual objects:

> Drawn from the real present object, the virtual object differs from it in its nature; it does not only lack something in relation to the real object it subtracts itself from; it lacks something in itself, by being half of itself, where the other half is posited as different and absent.
>
> (DRf, 135)

The treatment of Eros from within Deleuze's syntheses of time again shows the independence of his philosophy of time from psychoanalysis. Though he has taken the term of partial object from Lacan and others, it is sewn into Deleuze's own model of the syntheses of time. This allows Deleuze to define Eros in relation to the second synthesis and to memory through its function of attaching real objects to the pure past in virtual objects: 'This is the link of

Eros with Mnemosyne. Eros tears pure virtual objects from the past and makes them liveable' (DRf, 136). One final question remains unanswered, though. This concerns the relation between Eros and the attraction to the new, not to the second synthesis of time, but to the third. To understand this connection, we shall pass to Deleuze's study of death in relation to time and to eternal return in the following chapter.

5

Time and eternal return

ONLY DIFFERENCE RETURNS AND NEVER THE SAME

In *Difference and Repetition*, Gilles Deleuze's philosophy of time is constructed around three syntheses. Each one of these determines and depends on the others according to the concept of dimension. In the first synthesis of time, the present is the locus for the primary synthesis, understood as habit, and the past and future become dimensions of the present. In the second synthesis, the present and future become dimensions of a synthesis understood as the pure past. In the third synthesis, past and present become dimensions of the future. All these syntheses are processes, where process means a transformation of events and their relations. This is unlike a familiar sense of process taken from production, where a machine processes a material input and produces a product. In the syntheses of time the process is transformation across all parts such that any notion of independence is rendered obsolete. The machine is itself 'processed'. The syntheses of time are not processes by something, but in and involving something, for instance, a transformation of the past and the future in the living present when habit transforms a series of events.

Due to this interlocking of priority and dimensions, it makes no sense to treat the three syntheses of time as independent of one another. They are mutually dependent syntheses with different forms of determination on one another depending on which process is taken as prior. Deleuze has constructed a network of determinations and processes. This mutual dependence follows from the revolving nature of priority. The syntheses cannot be organised according to a hierarchy, because each one has a moment where it is prior to the others and because even when there is a relation of priority, this depends on other such relations.[1] Deleuze therefore presents us with a philosophy of time constructed around transcendental deductions of relations of determination which themselves depend on speculative moves, such as the affirmation of the new in the third synthesis of time. There is always a given presented speculatively, yet with much evidence drawn from philosophy, the arts and sciences, which then allows for the series of deductions.

In this chapter, one of the main aims will be to describe, analyse and criticise the role of an account of eternal return in the third synthesis of time. Here, eternal return will be explained and justified as process, rather than as an unchanging circle of time. As such, eternal return will be connected to the other processes that define the third synthesis: the caesura, assembly, order and seriation of time studied in the previous chapter and presented in chapter II of *Difference and Repetition.*

The critical tenor of this chapter will turn around a set of problems already raised during this study. First, what is the scope of Deleuze's philosophy of time and is this scope restricted through human or subject-based presuppositions? In this context, there will be a study of eternal return as a process defined partly in relation to death, or to a death drive (Thanatos) contrasted with a pleasure principle (Eros) as considered in the previous chapter. Second, what is the status of Deleuze's work on eternal return in terms of the truth and validity of his philosophy? This question will be approached from two angles: critically, through an enquiry into the robustness of that philosophy as a traditional claim to truth and validity; and critically and constructively, as the source of new definitions of what should stand as true and valid. All these questions come to the fore in the work on eternal return.

This delicate and yet crucial position of the work on eternal return is partly due to its strangeness as a key aspect of a philosophy of time in a modern era which has tended to view time as linear or, when it has considered time as circular, has done so on the basis of cosmology inspired by physics in a manner very distant from Deleuze. The difficulty of this new take on eternal return is also due to the innovative yet deeply paradoxical account of it found in Deleuze, one drawing inspiration from Nietzsche and from Plato, but which must nonetheless be viewed as very different from them, or as involving narrow and highly inventive interpretations of Nietzsche. Finally, the work on eternal return draws out critical problems in Deleuze due to its ubiquity and importance in *Difference and Repetition*. The idea of the return occurs at many points where the most original and difficult arguments of the book come together, notably in the closing passages of chapters II, IV and V, and in the conclusion.

I will start with one of the many statements of Deleuze's principle of eternal return from the conclusion to *Difference and Repetition*, a few pages from the end of the book. He sets the principle in italics as one of the last and most important principles to be stated or variously restated, according to a form of stylistic repetition reflecting the main tenets of the book: '*The same does not return, the similar does not return, but the Same is the return of that which returns, that is, of the Different, the similar is the return of that which returns, that is, of the Dissimilar*' (DRf, 384). The capitals on 'Same', 'Different' and 'Dissimilar' indicate the universal idea of each rather than a correspondence to particular cases or instances. Thus anything that is the same is the return of pure difference, rather than any identified difference. Any similarity is the return of pure difference.

Eternal return is therefore defined by the principle that only pure difference or difference in itself returns and never the same; where difference in itself corresponds to a concept of difference free of any prior dependence on identity or negation. The statement that the same is 'the return of that which returns' is therefore the core statement for Deleuze's philosophy of immanence as developed in relation to eternal return: nothing escapes the return of difference and there is no transcendent realm of the same. Difference in itself, as studied and defined in chapter I of *Difference and Repetition*, is not a difference between two identities or the negation of an identity. Were it defined in such a way, the principle would be tautological since the same would return with difference. The principle is not tautological in that way and another way of stating it is that every thing posited or working as the same presupposes that the same cannot return and that a difference that cannot be determined through the same has returned. *Any identified or comparable 'this' or 'that' will never return and has only appeared thanks to differences we cannot identify.* As a principle defining a process, this means that anything defined as the same thanks to an identity that it can be referred to or represented in, never returns. It also means that anything that can be defined as resembling something else does not return. Instead, the only thing that remains the same is the return of difference and the only thing that remains similar is the return of dissimilarity. So Deleuze's account of eternal return must be distinguished from any version that gives time as a circle or a return where what was before happens again, or returns in the same form, or recurs as resembling what came before. This stands in direct opposition to any notion of eternal return as rebirth, reincarnation, identical cycles, reminiscence of the same events, and the repetition of events, ideas or even patterns. Instead, pure difference, that is, all that was not represented and that which engulfed identity in earlier events, returns once more in new identities and representations in order to engulf them again. Nothing returns except difference in itself.

We have seen this work of pure difference twice already, in the first and second syntheses of time. The first synthesis of habit is not a repetition of the same events and things, but rather their passive transformation or metamorphosis. The condition for this metamorphosis as necessary process lies in the eternal return of difference. The second synthesis of the pure past also returns as pure difference, where a constant variation of relations in the pure past makes every present pass and pass as different from itself and from all other presents. Again, as necessary condition for this passing of presents we find the eternal return of difference. Without the third synthesis, Deleuze's first and second syntheses leave a set of questions unanswered and ill-explained. This is because both depend on identity and activity and both are open to an explanation in terms of a gradual filling of possibilities or a constant return of the same processes. The first synthesis of the living present depends upon, but is not reducible to, an activity in the present. Why is such activity not exhaustible, or limited to a given set of possibilities? It is because each such activity and any defined possibility is a one-off destined to pass and operating only through the return of difference. Each activity is incomparable even as it is necessary for the contracting of the past and the future. Any defined possibility is unique and has no final bearing on a future determination of possibility in its novelty and determination through pure difference. Why is the synthesis of the pure past an ongoing process and one that does not stop at a certain point or gradually slow down as fewer truly new living presents occur and are made to pass by the pure past? It is because the pure past returns eternally as a new and different future in every living present. Eternal return refuels the pure past and reignites the living present. The eternal return of difference underwrites the openness of Deleuze's system while explaining the necessity of identity as unrepeatable.[2]

A hammer pounds a hot rod of metal across an anvil. Each blow appears to make the same sound, to alter the molecular structure in the same way, to achieve a similar sharpness to the one achieved a thousand years before on an edge and an anvil a thousand miles away. According to Deleuze's account of eternal return, there is indeed sameness and similarity here, but it is in the way each blow, each anvil, each edge differs from all that came before and all that will come after because each one introduces novel differences through the return of pure difference. These are not only differences within each new stroke of the hammer, but rather differences in syntheses running back through every living present, through every past blow. They are differences in syntheses in the relations between the differences introduced in every past blow, in the whole of the past as pure past, or store of past differences. They are syntheses destined to return in the future differently within any new blow, new edge and new resounding clang. It is not that we cannot compare such blows or even the tipping point when a blow and reverberation trigger the return of an unbearable migraine. It is rather that when we do so we miss the complete process unless we also take account of the return of difference. It is not that we have to abandon regularity, patterns, statistical rules or probabilities. It is that these are only ever a partial representation of real processes and, if they are taken as the best or only way of relating to future events, they lead to a fundamental misunderstanding of the open nature of the future. The same and the similar are only the necessary media through which difference works and is expressed each time in a new drama encompassing all that has come before and all that will come later.

But is any of this true? And is any of this consistent and valid? Why should it be believed? And what is the status of these claims when compared with other philosophies of time and with other accounts of eternal return? If only difference returns, is there any identity through time at all? If there is no such identity, is there anything at all in any meaningful sense? Are you disappearing as you read this sentence and if you disappeared after you read the first word, who is reading the last one and how does it connect back to the first?

Does eternal return matter within Deleuze's philosophy, or could we simply bracket off these claims about the return of difference from those points where Deleuze makes more sense or runs in a consistent way alongside contemporary science? This last question at least can be easily answered. If we choose to discount the work on eternal return as in some way aberrant or unimportant, we will be severing off large parts of his philosophy, to be left with no doubt interesting and productive remarks, but with no consistent system and with many gaps and lacunae. We can see this through the interdependence of the syntheses of time. If we remove eternal return, then we also neuter Deleuze's accounts of habit as repetition and memory as repetition; in fact, we turn his novel definition of repetition into a mild conjecture mapped onto a series of observations. For an interpretation of Deleuze that seeks to give the strongest and most consistent version of his philosophy, there is no alternative to a full investigation and explanation of eternal return.

ETERNAL RETURN AND DEATH

Immediately before his last statements on the return of difference and on identities that do not return, Deleuze gives us a dramatic and inventive reading of Nietzsche's *Thus Spoke Zarathustra*. Here is one of the ways of expressing eternal return from Nietzsche's book: ' – and must we not return and run down that other lane out before us, down that long, terrible lane – must we not return eternally?' (Nietzsche, 1969: 179). This is one of the main texts where Nietzsche states his doctrine of eternal return, but I will not go into the detail of this reading; it has been studied well and extensively elsewhere.[3] Instead, I will study the conclusions Deleuze draws from his interpretation. The drama is significant because it draws out an implication of Deleuze's principle for eternal return that is hidden in its purest form, yet is a direct consequence of it. The principle of eternal return is not only about the new and a difference to be affirmed; it is about violence, death and the most difficult tests put to living beings. In order to explain these implications and to trace the great problems they raise for Deleuze's work, I will study the fifteen or so lines from the conclusion of *Difference and Repetition* where they are raised in their most stark outline.

The lines occur after Deleuze has drawn a problem raised by *Thus Spoke Zarathustra*. It is presented through three questions about Zarathustra's moods and his acts around eternal return (though Deleuze is already stretching this, since those events are not always closely linked to eternal return in Nietzsche's book). The three questions concern moments of anger, crisis, convalescence and silence. Together, they demonstrate that eternal return is not fully presented in that book, but is rather foreshadowed as an idea that can be misunderstood, that needs to be pushed further and that Zarathustra and his followers are not ready for. For Deleuze, eternal return would only have appeared fully in Nietzsche's account if an unfinished version of Zarathustra had appeared, as prefigured in Nietzsche's notes. In this version, Deleuze contends Zarathustra should have died. This association of death and eternal return will allow the third synthesis of time to appear in full: 'We only know that *Thus Spoke Zarathustra* is unfinished, that it should have had a follow on implicating Zarathustra's death: as a third time, as a third occasion' (DRf, 380). As we have already seen in his work on drama in the third synthesis of time, death and crisis are central to Deleuze's conception of that synthesis, but this can only be understood in full in the idea of eternal return.

Deleuze views Zarathustra's moments of anger in *Thus spoke Zarathustra* as a reaction to a possible misunderstanding of eternal return. Zarathustra fears and is angered by the idea that eternal return signifies the recurrence of the same events and world, or of all events and worlds. Not only is such a view self-contradictory, since it would fall foul of Leibniz's law on the indiscernibility of identicals (as studied in Chapter 2 here). No two cycles of return could be internally identical, for then there would be no sense of return in them due to the lack of any way of telling there had been a return. The idea of the eternal return of the same is also a view destined to lead to despair and anguish in the realisation that all efforts to escape evil and toil in the world are bound to fail in the eternal return of the same evil and same toil. This is why Deleuze insists on the perishing of the same and the return only of pure difference. The eternal return of the same is a great threat to Zarathustra and to Deleuze's belief in the future as open. Yet it is also why he raises further problems: 'The revelation that everything does not return, nor the Same, implies as much anguish as the belief in the return of the Same and of All, though it is a different anguish' (DRf, 381). This sentence marks a shift between two types of anguish and the transition into a concern about death. Where once there was anguish about the return of what we would not want to see again, now there is anguish at the passing of what we hold dear. An eternal monotony is replaced by an eternal passing away driven by the return of a violent novelty. It is violent because it makes anything identified, anything the same, pass away never to return.[4]

However, the use of the idea of anguish presents deeper problems. Does this imply a form of existentialism at the heart of Deleuze's account of eternal return, where the return becomes essentially a factor of existence conditioned by an affect?[5] Is the problem of such a return therefore the preserve only of beings capable of anguish? In such beings, is anguish to be taken as the fundamental affect or emotion? From the point of view of Deleuze's philosophy of time, and given the importance assigned to the third synthesis and to eternal return, affirmative answers to these questions would greatly limit the scope of Deleuze's philosophy, at a stroke making it irrelevant to anything incapable of anguish. It would set stones and insects outside the ambit of time, banished to the first two syntheses but incapable of living the third. It would also set beings beset by anguish at the centre of time and at the point of a hierarchy, perhaps one of despair where the top is not a pleasant place to be, but a ladder for categories of beings in time nonetheless. Even more seriously, perhaps, a fundamental role for human anguish would greatly restrict the processes of habit and passing away into memory of the first and second syntheses, for their dependence on the new would be shorn of its support from the idea of eternal return, restricting true novelty to humans.

A first response to these criticisms can be found in the claim that the appeal to anguish occurs only in the context of Deleuze's reading of eternal return in *Zarathustra*. Anguish is then not a necessary sign of eternal return or experience of it, but rather a result of eternal return for a limited set of beings. Eternal return holds for all beings, or more properly, for all processes; it is expressed in beings as anguish only in beings capable of such expression. The principle of eternal return for Deleuze is then not deduced from anguish, nor is the experience of anguish a necessary condition for being situated in time in relation to eternal return and, hence, in relation to Deleuze's third synthesis of time. Deleuze's philosophy is then not constructed on existentialist experiences or terms, but it allows for explanations of affects such as anguish in relation to much wider processes, such as the eternal return of difference and the perishing of the same and of similarity. We can see here how the broad speculative moves in Deleuze's metaphysics are important for detaching it from a narrower experiential ground. It is essential for him to present eternal return as a formal principle, rather than as a human existential experience, or as the property of a specific type of thought, or as signified by a particular affect.[6]

Yet, even if we accept these answers, Deleuze's account continues in a vein appearing to restrict and rarefy his philosophy: 'To conceive of eternal return as the selecting thought, and repetition in eternal return as the selecting being, is the highest test' (DRf, 381). If we set aside anguish as the result of this conception, we are still faced with the objection that now eternal return seems to depend on a capacity to conceive of it and then to react to it as a test. So now a form of conceptualisation seems to be a condition for undergoing or experiencing eternal return. This is in many ways even more restricting than the earlier conditions, since even fewer beings will be able to conceive of such a thought. This would imply that the test is one for a select few. Again, an answer to this objection can be formulated according to a distinction drawn between the necessity of eternal return and different ways in which it is received. Eternal return does not depend on a conscious reception and concept of it, but rather Deleuze means that if thinkers are to understand a process applying to all other processes, they should understand it as the eternal return of difference. The understanding and concept are not preconditions for the return, but rather a certain concept of return is necessary for its adequate understanding.

When it is understood in that way, there is a corresponding comprehension of eternal return as the highest test. It is always such a test for all beings, but cannot be conceptualised as such until it is associated with the concept of the non-return of the same and eternal return of difference. As such it is the highest test within any living present, since it dictates the passing of all its identities, of everything that is the same in and around it, and only the return of that which differs as an expression of pure difference. For self-conscious beings the test then becomes a double challenge: how to bear the passing of its identity and how to affirm the return of difference. All processes in the living present evolve within the constraints of the non-return of the same and the return of pure difference, though not all can reflect on it. The evolution of plant species takes place within those constraints. So does the short life of a fruit fly. As does a pebble polished slowly over millennia by the tides. This means that the concept of selection used by Deleuze, in the sentence following the one on the conception of eternal return, is not reserved for beings conscious of such selection. Everything exists in time according to the three syntheses, but how these can be received is not the same for all things, in particular, in relation to a capacity to conceptualise and act consciously in relation to the syntheses. This does not imply a higher position or capacity for beings with those capacities, given their dependence on passive syntheses outside the grasp of conscious activity.

Deleuze's next sentence refers back to his work on the third synthesis of time in chapter II of *Difference and Repetition*. It introduces the idea of time out of joint again and also begins to resolve a puzzle in his definition of the third synthesis when it is presented as cut, assembly, order and seriation of time, yet also as a circle, an eternal return. How can time have an order of before and after, set into an irreversible and asymmetrical series according to caesuras, yet also be circular? If it is an eternal return, then does this not imply that what was after can become before, if we compare successive cycles with one another? The answer is that *the third synthesis of time is both an irreversible series and a cyclical return.* It is an order for everything remaining the same, where nothing can return, but it is a cycle for pure difference, the only thing returning each time. This duality of time leads to its violence and perhaps cruelty towards those wishing to preserve an identity against the passing of time: '[. . .] time set in a straight line pitilessly eliminating all those engaged on it, who appear on the scene, but appear only once' (DRf, 381). This elimination explains the nature of the test of selection, for instance, for conscious beings: how to live the inevitable passing away in time in such a way as to also participate in the return of pure difference? Deleuze is harsh in his prescriptions in answering the question: '[. . .] those who repeat identically will be eliminated' (DRf, 381). This is a law addressed not only to those capable of acting upon it after reflection – and even then the reflection can only be partial; the law applies to any process.

So when Deleuze restates his principle of eternal return in terms closer to death, he is not addressing a subset of mortals of whatever kind (anguish-ridden, possessing a power of reasoning, capable of conscious action, animal). His statement applies to all beings and the pronoun in his sentence should not be read as standing in for human actors, but rather for everything, that is, for all processes: 'Not only does eternal return not make everything return, but it makes those who cannot bear the test perish' (DRf, 382). It is at this point that we can revisit a difficulty with terminology in interpreting Deleuze, as well as drawing out some of the distinguishing features and possible weaknesses of my interpretation and choice of language. There is a general difficulty in his argument in relation to language, if language is taken to depend on the identification of a concept or meaning, since any being associated with such an identity must perish and is not directly expressive of pure difference. In the current argument this comes out strongly in the choice of words such as 'thing' and 'being' that we normally associate with an identified object. Any such identification is illusory when viewed according to Deleuze's account of time, since the syntheses of time all refer to ongoing processes against which any 'thing' or 'being' is an incomplete and misleading representation. Even in the distinction drawn between sameness and pure difference in the test of eternal return, I claim that the difference is between different kinds of process: those working towards an identity and those working in processes of becoming and change, novelty and transformation. So Deleuze's concern for perishing is about processes resistant to difference, processes focused on resisting change in the name of identities. It is justifiable to continue to use terms such as 'thing' and 'being', since identity is still a component of his philosophy, but it is essential to keep in mind that any such identity is fleeting and only one side of a reality more deeply conditioned by pure difference.

It is partly on these grounds that we can find an answer to the critical point ascribing a privileging of certain life forms to Deleuze's work on death, perishing and eternal return. He provides the detail of this answer in chapter II of *Difference and Repetition* in a critique of Freud's work on death and a wider critique of any definition of death as a return of living beings to undifferentiated and inanimate matter. This is another contrast to be made between Deleuze's version of eternal return and others. Eternal return is not a cycle of death and rebirths through cold matter by souls or consciousness-bearing life forms.[7] For him, eternal return is almost exactly the opposite: it is the passing away of that which is inanimate in sameness and identity and the eternal return of multiple forms of difference, where sameness is a necessary fate for any process in the passing away of the living present, but where the return of difference is only given to those processes avoiding the fixity of an identity. Death then is not an objective state that living things fall into and humans dread: 'Death does not appear in the objective model of an indifferent inanimate matter, to which life would "return"; death is present in the living, as a subjective and differentiated experience endowed with a prototype' (DRf, 148; DRe, 112). Death is in the pure form of time as it makes all forms of the same perish. This stone, this fly, this man, this pyramid, this state are all necessarily perishing. This explains why Deleuze counters any definition of death as negation or opposition: death is not against life, it is in life. Death isn't non-being, it is being: 'It is neither the limitation of mortal life by matter, nor an opposition of an immortal life with matter, that endow death with its prototype' (DRf, 148).

This prototype for death is given as a set of problematic questions such as 'When?' and 'Where?' It implies that death is always something unknown and to come: a problem regarding the future. It must be understood in relation to Deleuze's work on eternal return as test. Death is not only about the passing of identity, but also about the problem of how things live on through the eternal return of differences that make their and other identities pass. Evolution and change are a form of death, not only in the passing of things and creatures that will never return, but also in the moving towards new processes through difference in itself. So there are two deaths: '[...] a personal one that concerns the "I" and the self [...]' and another 'strangely impersonal one, with no connection to the "self", neither present nor past, but always to come, source of an unceasing multiple adventure in a question that persists' (DRf, 148). This adventure is not to be associated with human subjects, but rather with series of processes that draw the new and difference back into two processes of dying. This is why there is a double death in Deleuze: death as subject and death as difference affirming process.

The dual aspect of death allows Deleuze to give a definition of eternal return in relation to death with no reference to an essentially subjective or objective understanding of death:

> If eternal return has an essential relation to death, it is because it promotes and implies the death of everything that is one 'once and for all'. If it has an essential relation to the future, it is because the future is the deployment and explication of the multiple, of the different, of the fortuitous, for themselves and 'for all times'. Repetition in eternal return excludes two determinations: the Same or the identity of a subordinating concept, and the negative of the condition that would relate the repeated to the Same and would ensure the subordination.
> (DRf, 152; DRe, 115)

This pure form of Deleuze's principle of eternal return has universal applicability. Anything that is not extended into the multiple processes rendering it different, anything standing as one, does not only perish, but perishes never to return. The multiple ways in which things become different will return eternally. This principle ensures the openness of the future as the third synthesis of time and defines death in relation to it, not through any particular experience of death, but rather as a two-fold implication of the process in the passing away once and for all of the same and the becoming eternally of difference.

SERIES AND ETERNAL RETURN

Deleuze's use of the phrases '*une fois pour toutes*' (once and for all) and '*pour toutes les fois*' (for all times) in his purest definition of eternal return draws out two serious difficulties but also the originality of his work on time in relation to the future. The main difficulty is drawn out well by a translation problem, since the English version of '*pour toutes les fois*' adds a concept of time not present in the French. How can eternal return apply once and for all and for all times? Does not any process take time and therefore invalidate the claim that eternal return is once for all times? Is eternal return not only immediate, changing all times in an instant, but also immediate for all events, changing everything in that instant? If that is the case, is Deleuze not committed to a philosophy of time with a component that is outside time, either in a timeless understanding of human or world time that stands outside them, or in a succession of independent instantaneous states of everything? *Is eternal return the time of a transcendent omniscient God outside the process of the world, or is it the world but set in successive instants, as if lit by a stroboscopic lamp?*

It is helpful to separate these questions into two. First, why is eternal return each time for all events? Second, why does eternal return not take time? We have already seen versions of the first question and answers to it in relation to the second synthesis of time and the pure past. Each passing present must pass into the whole of the past because each passing present is made to pass by all of the past and not part of it (otherwise the past will be made of separate independent parts). We have also seen versions of these arguments in relation to the third synthesis which is a cut, assembly, ordering and seriation of all of time. The addition of eternal return is an explanation of how these two syntheses of time fit together. The pure past is strictly differential, in the sense of made of the differences that return in eternal return. With the new all these differences return in a novel way in relation to a passing present that will not return. The cut in time takes place with that passing present. The assembly is in the transformation of the pure past. The ordering is in the non-return of any passing present as the same. The seriation is in the difference between the pure past and its return as a novel difference. Eternal return is once and for all, because it transforms all of the past in relation to the future.

This explains Deleuze's at first sight strange claim that a move in a game associated with eternal return affirms the whole of chance.[8] What he means is that a novel move affirms all differences (the whole of the past as pure past), goes beyond all known moves (each move as the same move) through a move that is necessarily victorious (one that returns eternally as difference, but never as the same).

Here, chance does not mean probability, it means pure difference or novelty, a move beyond established probabilities:

> Not restrictive or limiting affirmations, but ones coextensive to the questions asked and to the decisions from which these emanate: such a game carries the repetition of the necessarily winning move, because it only wins by embracing all combinations and all possible rules in the system of its own return.
>
> (DRf, 152; DRe, 116)

Any player is both a winner and a loser in this game. It loses as 'one and the same', as an identity that can be conceptualised and that never returns. It also always wins, in its difference and novelty. Survival will be a balance between these two necessities, since we cannot have one without the other. To affirm difference we have to be the same, but in being the same we pass away unless we participate in the affirmation of novel differences. This leads to the conclusion that anything not in a process of becoming is dead, in the sense of incapable of return, whereas anything becoming is also dying, but as such also living on differently through the return of pure difference. There is an eternal death, then, in the sense of the perishing of the same. If eternal life is defined as an eternity within a defined identity, then it too is really death, since eternal life is only open to that which participates in or expresses pure differences in becoming. These are then two deaths from *Difference and Repetition*: a general death, once and for all, of the same; and a singular dying through multiple ways of becoming, a dying that is also a returning differently.

The questions about the time eternal return takes are therefore not phrased quite right, since they allow for a quick answer. From the point of view of Deleuze's philosophy of time, it is unimportant whether something takes place in an instant, because all time is process and the concept of a point-like instant comes from a view of time explicitly rejected by Deleuze's concept of synthesis. Similarly, the idea that it is a problem for something to happen across all events at the same time is not in itself a problem, since again, there is no assumption in Deleuze's philosophy that processes should be measured by an external time and one that posits that processes should be measurable as taking a stretch of time above zero. The deep problem for Deleuze is different. It lies in explaining how his definition of eternal return accords with his account of all the other syntheses of time. How can we relate the contractions of the first synthesis of time to eternal return? How can we relate the synthesis of all of the past in the second synthesis of time to the return of difference in the third synthesis? In other words, given the account of different dimensions of time dependent on which one is accorded priority, how do all these dimensions fit with one another? Or are the dimensions presentations of views of time that must be considered completely independent of one another and, if they are, is Deleuze's philosophy of time really three separate philosophies of time?

Deleuze answers these questions thanks to the concept of series, both in the conclusion to *Difference and Repetition* and in chapter II. This is significant because it shows the connection of that book to the contemporaneous *Logic of Sense*, where the concept of series dominates and is developed to its full. It is tempting to see the latter book as more impressionistic and experimental, more poetic, than the former, but in fact the two books interact and inform one another theoretically as well as through their different styles and approaches. In the conclusion to *Difference and Repetition*, Deleuze introduces the answer to the question of the coherence of the different syntheses of time through the concept of series by asking himself a critical question: 'What, however, is the content of this third time, of this formlessness at the end of the form of time, of this decentred circle displacing itself at the end of the straight line?' (DRf, 382; DRe, 299). Eternal return is a decentred circle because it has no ideal or ground to return to; it is nothing but a differential relating the before and after of the straight line described by the caesura and order of the third synthesis, the form of time. It is formless because it allows of no identity, of no sameness whatever. The question about content is then about how it works as process. How does eternal return work in relation to the other syntheses, when these have been described in practical and identifiable ways as contraction and synthesis? How does eternal return work in relation to the other ways of describing the third synthesis of time, as cut, assembly, ordering and seriation?

The answer is that the content of eternal return is series and simulacra.[9] Eternal return works by setting off differences within series thanks to simulacra. Put more simply, it means that eternal return works by relating differences to each other in series. For instance, in the contraction of a series of events in the living present, eternal return explains how this contraction is a metamorphosis of all that it contracts through the introduction of differences carried along the series by simulacra. These simulacra are not identified objects, but rather, in the language of *Logic of Sense*, empty places and placeless occupants, that is, things working within something else, either as a place for something absent or as a thing but with no assignable place. The simulacra are therefore not identifiable within the series. Classic examples of these processes and simulacra would be an unidentified culprit in an investigation, or the source of a disturbance within a system with an unidentified series of effects. The culprit is an empty place we seek to fill during the investigation and, as we search, we alter the relations of all series dependent on where and who we assign to the place (for instance, in the way all the relations of history can be made to vary when we alter the culprits for events; *the CIA did it, it was a Russian double-agent, it was a rogue operator*). The source of a disturbance, a new rise in popular anger at brazen greed from its leaders, for instance, has no assigned place but it runs along all the places it can take, altering their relations (*only now do we understand the collective will of subjugated people evolving over centuries*).

This appeal to simulacra raises a possible misunderstanding and an associated criticism. I want to phrase these in relation to Deleuze's philosophy of time, though they are related to a more general version applicable to any instance where he relies on the concept of simulacra and to wider debates about the nature of the real in his work. The misunderstanding would be that he is opposing simulacra to another kind of real entity in his system. This is not the case: simulacra and differences are all there is in series and an error is made when something that is a simulacrum is taken to be objectively real. A similar error is made when a pure difference is taken as objectively real. Difference is a potential in Deleuze's philosophy when it is taken as pure; it is part of a real process of differentiation when it is expressed in syntheses that are both actual and virtual in relations of reciprocal determination. However, this invites the objection that there are real objects and that these correspond to correctly identified places or occupants within series. So we can speak of the real murderer and the real effects of anger once these have been recognised, most likely on some objective scientific basis. Real time would be the time that corresponds to this basis. Thus, for instance, we could say that until we find the causes and effects of climate change we are dealing with simulacra, but once there is reasonable scientific evidence for these, then we have the real objects and we can speak of the time they occur in. This would not be the time of Deleuze's many syntheses, since these explicitly deny identifiable objects by insisting on processes requiring simulacra and pure difference. We must make such identifications, and we should do so accurately, but we should never view them as complete or as the final version of the real.

Deleuze has an answer to the objection about objectivity through his work on syntheses. It can be readily agreed that murderers can be identified and effects recognised; however, what will not be accepted is the claim that the real should be associated with the objective entities since they are but abstractions from the synthetic processes underlying them. So it would be a mistake to speak of the real time of objects, because when viewed in relation to the processes giving rise to them, the objects are not real. They are less real than simulacra, since the abstraction required to recognise them as real is a false representation of the objects, the syntheses of time and the forms of representation at work in constructing them. Real objective time does not exist. It is a product of abstraction and representation that are open to a critical analysis in terms of process time. This critical side to Deleuze's philosophy has perhaps been underplayed here, due to my aim of explaining and criticising his philosophy of time and the speculative, transcendental and empirical methods used to deduce and construct it. However, one of the important results and minor aspect of the construction lies in the critique of falsifying forms of abstraction and representation, notably in the philosophy of time.

In *Difference and Repetition*, Deleuze defines simulacra through specific applications: 'Simulacra essentially implicate, under a same power, the object = x in the unconscious, the word = x in language, the action = x in history' (DRf, 382). Eternal return puts something to work, but this thing is never captured by any of its representations, whether it is a conscious representation, a signification associated with a word, or the subject and object of an action in history. Instead, it is deduced from its effects and these effects are relations of differences: 'Simulacra are those systems where the different is related to the different through difference itself. The essential thing is that we find no prior identity and no interior resemblance in these systems' (DRf, 383). So when Deleuze describes eternal return as a process that operates on all things at the same time, he means that differences work through all series through simulacra and that the differences occur throughout the system at the same time. This working through destroys all identities associated with the series, since they disappear with the changing of relations of difference. All differential relations, however, return in the new relations of difference running through the system. The transformation operates in the living present and in relation to the pure past such that every synthesis of time is transformed with eternal return as the principle of the future. The future as the new and the new as pure difference changes everything, all the time, everywhere.[10] All relations of comparison or resemblance are also destroyed in this process, because the novel relations of difference set up new systems of differences, invalidating earlier comparisons and resemblances. When the present and the past are considered as dimensions of the future their moments of sameness are cancelled. However, does this imply that there is nothing the same in Deleuze's series and, if so, does he fall into a problem raised by Heraclitus's paradox? In stating that we never step into the same river twice we presuppose some way of identifying the river as the same.

DISPARITY AND DIFFERENCE IN ETERNAL RETURN

Deleuze tackles the objection that sameness is a presupposition of eternal return, understood as the work of difference through simulacra in systems of series, in the penultimate section of chapter II of *Difference and Repetition*. He begins by distinguishing two formulae: one where resemblance and an identical concept are the conditions for difference and another where 'resemblance, identity, analogy and opposition can only be considered as effects, the products of a first difference or a first system of differences' (DRf, 154). In order to defend the second formula where difference is primary, Deleuze claims that systems are made of coupled heterogeneous series. So a difference does not only work along one series, but it is related to another difference working along a different series. This means that in systems, series are related not through a difference related to similar elements, but rather through a relation between two differences. Deleuze then associates this with three sub-processes encountered, for example, in physical systems: a coupling between heterogeneous series, internal resonance within the system and a forced movement that goes beyond the initial coupled series. So, for instance, there could be an increase in an itch along a series of mental irritations coupled with an increase in scratching along a bodily series of nail against skin. These could lead to a resonance where the scratching increases the irritation and the irritation the scratching, such that the rise in waves of irritation overflows into other series, from a localised itch and irritation to one all over the body or associated with a different series entirely (with mental image or a fantasy, for instance).

Deleuze then claims that this establishes a dynamic system of

Time and eternal return

resonances based on differences.[11] This dynamic system would then explain how eternal return works as difference carried by simulacra through series in response to an objection claiming that resemblance and identity are preconditions for the operation of difference. However, in his familiar pattern of self-critical questions, Deleuze enquires about the limits that might be necessary for such an account to work and about how such limits might be set. This is a doubly difficult objection for him to consider because it reintroduces resemblance and brings up the possibility of limitations on series, both of which counter his view that eternal return works everywhere at the same time through pure difference. Can any series resonate with any other or should they have a prior resemblance? Do such resemblances allow for us to classify series and divide them according to whether they can be connected or not, thereby dividing being into separate natural kinds? In answering these questions Deleuze returns to one of the most difficult terms from his process philosophy of time: dark precursors.

The dark precursor is a speculative term in Deleuze's philosophy.[12] It is introduced and exemplified by thin accounts of natural events, such as a lightning strike, but these are not fully developed or defended; they are a source of ideas and mild confirmation, but in no way constitute a demonstration of the validity of using such a term. Further evidence for this can be found in the variety of reference points, where a discussion of literary series from Raymond Roussel, Lewis Carroll, Joyce and Proust far outnumber much more scant discussion of natural events. It is certainly possible to seek to provide such a valid scientific or perhaps mathematical basis, but it is neither present in Deleuze's work nor in my view called for. This is because Deleuze's argument is quite different from any philosophical naturalism. He is seeking to show that, on the basis of the speculative introduction of the term, he can explain relations between series through pure difference without having to appeal to concepts of identity or resemblance. The argument is therefore not empirical but rather about philosophical consistency and explanatory power. According to Deleuze's argument, it is both more consistent and of better explanatory power to assume that difference is prior in the relation to series and requires no assumption of resemblance or identity. In order to show this he gives the following definition of the dark precursor:

> Given two heterogeneous systems, two series of differences, the precursor acts as differenciator of those differences This is how it puts them

immediately into relation through its own power: it is the in itself of difference or the 'differently different', that is difference to the second degree, difference with itself relating different to different through itself.

(DRf, 157; DRe, 119)

The dark precursor therefore explains how differences between series are related not through a principle of resemblance and through the identity of that which connects them, but rather through a third term that itself has no fixed identity because it holds differences within itself. The dark precursor is then another simulacrum. It is invisible, in the sense of not having a represented identity, and only appears after it has done its work, after the series have been related: '[. . .] it is precisely the object = x, the one that "is missing in its place" and missing its own identity' (DRf, 157). This reverses the requirement for resemblance between series, making resemblance only an after-effect of their prior relation through difference: 'Identity and resemblance would then only be inevitable illusions, that is, concepts of reflection that would account for our inveterate habit of thinking of difference through the categories of representation' (DRf, 157).

There is then no necessity to thinking of the relation of series in eternal return in terms of resemblance and identity. Deleuze's syntheses of time are constructed in such a way as to reverse the preconceptions in favour of identity; preconceptions that are illegitimately imposed on experience and experiments. This allows us to respond to one of the main objections to Deleuze's philosophy of time and in particular to his account of eternal return in terms of difference. Identity is not a necessary condition for the presentation of difference or for the definition of difference as process, because it is possible to define difference and processes based on difference differentially. The dark precursor is a way of explaining how differences appear in relation with no requirement to identify how they differ or how they resemble. Deleuze gives a further definition of the dark precursor to underline this point. The dark precursor is 'disparate' ('*dispars*'). It is itself in a process of continual change in relation to the series it relates and in relation to where it relates them: 'In each case its displacement space and its process of disguise determine the relative size of the related differences' (DRf, 157–8). It is important not to read this statement as implying an external measure of size, which would introduce resemblance and identity again. The key term of the statement is 'determines', which must itself be understood in relation to the definition of the dark precursor as the agent of difference. There is no relative size of the related difference or series outside their relation through a dark precursor: 'For the in itself, the important thing is that, whether small or large, the difference should be internal' (DRf, 158). When the precursor determines a size, this only means that the process relates different internal scales of differences, for instance, in terms of a large resonance between series. *It may have been a tiny, almost irrelevant habit, but their relation shattered itself on it. As the winter storms drifted south, the butterflies gradually disappeared.*

The internal differences are essential for understanding Deleuze's counter to the objection that series are successive in time, that is, that events occur sequentially along them. A corrupt building permit in 1967 leads to a high-rise constructed with substandard concrete in 1968. The harsh winters of 1982 and 1983 then weaken the concrete. A small earth tremor brings the top layer down in 1984. Second after second, each successive floor of the building slumps down on to a lower one, finally crushing everything between them. It seems clear that these events and processes happen one after the other, and Deleuze does not have the ready counter that with perfect knowledge of the laws governing the unfolding we can consider it to happen independent of time, since once the first fact is registered all others follow necessarily. The response from within Deleuze's philosophy is first to focus not on the chain going from past to present, but rather on the present in each part of the chain and on problems at that point, so the question is not about what caused something in the present, but rather what we should do given a problem in the present in relation to the past and to the future.[13] So it is not so much the question 'Why did this building fall down?' though this is a question relevant to drawing up the problem. It is rather that the problem is made up of many questions: 'How should we build now?', 'Who should build now?', 'Where should we construct new buildings?' Each new answer and practice in response to these questions sets off differences in series running back in time and demanding a review of the past, as a well as a demand not to repeat it. Each novel event also sets off chains of differences through series. So it is not that we need to deny causal explanations, it is rather that they are insufficient. The work on eternal return allows Deleuze to demonstrate this insufficiency by connecting the new and radically novel events, to the necessary passage of the same and to the demand for experimentation in relation to dark precursors, that is, to signs both of the failure of the same and of the event of a novel difference.

6
Time in Logic of Sense

OF WOUNDS AND TIME

The tourniquet is not working. The red stain on the bandage is growing wider and blood is now dripping at its edges. Nothing more can be done. The life is lost and is draining away in short shallow breaths, soon to close, with an end you know nothing of, save your rising dread as it passes from vague suppressed images to overwhelming physical panic and incapacitated thought. Time here is a peak and an onrushing void, not a dark unknown space but rather an obliterating spread, inching out along uneven channels from a summit of pain and injury, slowly fixing the past and erasing the future.

When time is defined as process, agony can become intrinsic to it. Neutral time, where each instant is equal to any other, or mathematical time ruled by abstract equations, fall short of a lived agony and a witnessed dying. Even if they are mapped from relative spheres and according to multiple vectors, the peaks drawing all lines towards them are still too generic, too removed from individual experience of physical privacy and isolation, where solitude is not a detachment but rather a dramatic private focus, like those waking intervals after a nightmare featuring our own children. All is there, but nothing is at it should be, or as we need it to be. All is there, but its assumed projection forward is replaced by a premonition of the temporary stasis of grief and its longer-lasting aftershocks.

When time is defined as process, agony also becomes responsibility. Affect neutral theories of time and of space find it hard to explain the singular focus of feeling and attention, let alone rise to the duty to show a way through it. The problem is not with the general affect, which can be observed and registered; it is how it is mine, or ours, or theirs and locked away from any general representation. Individuality determined by affect and process resists the implicit comparability afforded by generality. Beyond a failure for others, the silence of general abstraction is also a reflexive curse: it is to be able to think deeply but still to miss the joy and the pain of lived experience.[1] Yet a different set of curses might befall time as manifold processes. Must we have a time for each wound and each agony, trapped in its own private pain? Or, perhaps, each wound is shared but only by a subgroup of minds and bodies capable of such experiences? Few thoughtful witnesses to animal distress can easily accept that true mortality is reserved for humans alone, and even then not for those humans rendered merely animal by some lack or absence. If there is such truth about time as process, in experienced dying, for instance, is all of life then rendered gloomy and morbid, reminded of its necessary participation in perishing and of a final end almost impossible to live well?

Deleuze's process philosophy of time, constructed thanks to his transcendental and speculative moves, has many resources not only for thinking about the singularity of times but also for expressing and dramatising their most difficult events. Here are some of the processes determining time as individual through singular events: the concentration of the first synthesis of time in the living present; the passing away of that present into a dynamic and intense past; the rupture of time, then its assembly, ordering and seriation as irreversible and asymmetrical; the return only of pure difference and the eternal perishing of the same in this rupture torn by the new. All of these syntheses of time provide a system of processes capable of articulating and expressing singular dramatic events. Deleuze's difficulty is in the other hard questions around ontological exceptionality. The human is in danger of becoming a special existence, at the expense of other living and inert processes (which perhaps therefore risk even greater danger due to their comparative expendability). This dual threat is compounded by the narrow justification of that special place, since human life becomes articulated not through its richness and openness of ideas, creative powers and affects but only according to those thoughts and feelings characteristic of mortality and being towards death.

The reading of Deleuze's philosophy of time given here has already frequently encountered these problems. The living present in the first synthesis of time raised the danger of an atomisation of processes around each living present. It then came up against the deeper problem of an infinite division of each atom into smaller and smaller lives organised around minimal concentrated series and presents, to the point where the living being was reduced to minute events of life, requiring a subsequent aggregation to counterbalance the prior differentiation. Memory too presented great difficulties around the question of where to situate the individual and how to relate it to the pure past. Actual memories disappear into the pure past of the second synthesis, such that any memory is a false representation not only of the pure past, but of any past present. In relation to the pure past it appears that the individual is both nothing, as that which is sucked into a depersonalised dynamic process of differential relations, and everything, as the condition for the pure past alone, its actual fuel and ongoing generator of novel dynamisms. Deleuze's account of the third synthesis of time rested on individual dramas, on Oedipus as translated by Hölderlin and Shakespeare's Hamlet, whose experience of time out of joint provided Deleuze with the step taking him from a caesura in time, to its assembly, to its ordering into infinitely many series of before and after, and to the difference between the time before the cut and the time after it.

All the processes of the first and second syntheses themselves depended on a stranger condition for the novelty in each one of them; but it also depended on them. Eternal return, as the condition of the new in other syntheses, underwrote the return of pure difference alone, thus also guaranteeing that the same and identity could never return. Yet this return depended on the pure past as reserve of difference, avoiding the need for creation out of nothing. It also depended on the same and on the living present as supports for the return of difference; a sacrifice of the same in repetition for the sake of difference. This account was also built on individuation and affect through the figure of Zarathustra and his angst. The apparent requirement for the special individual was not limited to Nietzsche's seer. It ranged more widely to beings capable of angst and of conscious awareness of sameness and return, not only at the thought of the tedium of the return of the same, but also in the realisation that everything must perish. Even Deleuze's solution to the elevation of Zarathustra, through his death, focused eternal return on death and morbidity, still making it a thought for reflection on finitude and accompanying affects of despair and thwarted hope.

Deleuze's philosophy allows for a response to these critical points, not through the elimination of the individual and of affects,

Time in Logic of Sense

but through their extension. 'Individual' does not token humans or animals in time, but rather an individual path of syntheses, a form of determinacy against an undifferentiated background. It is not an identified individual of whatever natural kind or metaphysical sort, but rather a process of individuation. Taken to the limit, the background for this individuation is an unmanageable chaos. Individuation is therefore a necessary condition for determinacy in Deleuze's system. It is a selection indicated by a difference carrying along multiple series. This selection is at work in the living present, in the individual difference of a grain of sand or rice, or in the timorousness of an individual mouse. This does not lead to an atomism, because all the series differentiate one another, they resonate, not only along the syntheses of the living present, but into the pure past and through the eternal return of difference in each individual. We therefore have a time of the individual as three syntheses, but this time is also the time of all such multiplicities. Time is not a static One, then, but a mobile multiplicity in which every new event, that is, every event, not only takes its place but also leads to a disturbance and harmony through the novel differences it introduces. These differences are the condition for its determinacy as individual.[2]

Yet has this inclusiveness of individuals and series gone too far away from human affects? Does it lead to an indifference to what is special in the human? Is it incapable of staging a moral philosophy of time where decisions for some syntheses and for some affects are elevated above others? Or are we to weep over a stain on the surface of a marble statue as much as we would over a spoiled dressing covering a mortal wound?[3] To answer such questions and to enquire more deeply into problems of time and morals in Deleuze's philosophy, it is helpful to turn to the other major site for his philosophy of time: *Logic of Sense*. The book provides many striking technical contrasts in vocabulary and structure, and alternatives of context and examples, with the treatment of time in *Difference and Repetition*. Nonetheless, it will be argued here that there is great benefit in reading them together, while avoiding a final conflation of the two texts. Like Deleuze's series and events, they resonate with one another and enrich each other, but they also introduce tensions. Sometimes, these opposing pulls allow for dramatic and difficult conversations between the two books. This is the case around the question of the position of wounds and different registers of suffering in Deleuze's philosophy of time.

The approach I will take to *Logic of Sense* will consider some of the more obvious theoretical contrasts with *Difference and Repetition*. It will also move through a number of the central questions already posed in relation to the three syntheses of time. The aim, though, is not to stick to these theoretical issues in themselves, but rather to reflect on whether they bring interesting divergences and common foci to bear on the deeper problems of Deleuze's approach to time. Perhaps the most glaring difference between the two books in relation to time lies in the contrast between the three syntheses of *Difference and Repetition* and the two times of *Logic of Sense*, Chronos and Aiôn. The names for these times are taken from the Stoics, a central philosophical movement for *Logic of Sense* but a minor one in *Difference and Repetition*. I will argue later that the two times lead to the same six relations as the three syntheses: the past and the future as dimensions of the present, in Chronos; the past and the present as dimensions of the future, in Aiôn; and the present and the future as dimensions of the past, in the relation between Aiôn and Chronos, as mediated through intensity.[4] This dimensional aspect of the relations between times is important for both books and it therefore raises the problem of priority for *Logic of Sense*, as it did for *Difference and Repetition*. This problem would not come to the fore through a possible priority of eternal return, or the pure past, or the living present, but rather through a priority to be ascribed either to Chronos or to Aiôn. However, behind such questions of priority in processes, the main problems in this chapter will be moral. Can Deleuze's philosophy of time provide a basis for moral distinctions and action? Is that even one of its aims or one of its areas of concern?

HOW MUCH, HOW AND WHERE?

The philosophy of time appears early on in *Logic of Sense*, in its first and second series of paradoxes 'of pure becoming' and 'of the effects of surface'. However, the book is not a straightforward unfolding of arguments and topics, and it would be a mistake to put too much emphasis on whether concepts and ideas appear early or late in the book. Context and juxtaposition are far more important than order for the book and for each series, so it is best to study the topical surroundings, arguments and concerns for each consideration of time. In the first series, time is introduced through a reference to Plato, an important source and sounding block that we have already observed at work in *Difference and Repetition* according to two contrasts, between Plato and Deleuze's ideas of the circle of time and between their treatment of originals, pretenders and simulacra. It is the problem of simulacra that comes to the fore in the first series. It does so in a triple context: through the question of a time that avoids the present; through the technical problem of pure becoming; and through the moral problem of the role of common sense and good sense in relation to multiple and contradictory directions of becoming. These problems are in fact connected and inseparable. They generate one another. However, given the difficulty of each, I will begin by treating them apart and then draw them together.

The first statement about time in the first series is about simultaneity and pure becoming: 'But it is at the same time, in the same play, that one becomes bigger than one was, and that one renders oneself smaller than one becomes' (LoSf, 9). It is important to resist a first misreading here. It is common to think that 'at the same time' means 'in the same instant' and 'not at different times'. Neither of these definitions is correct here. We can find a first clue as to why this might be the case in the second clause of the statement. Deleuze is not referring to instants but to plays, to a process like a move in a game. These do not take place in an instant, like the crossing of a threshold on a line, but rather spread through a structure and system immediately. His point is therefore that all processes of becoming take place together and take no time to do so, that is, they neither occur in a specific instant nor over a measurable stretch of time.

The use of 'at the same time' in the first series is therefore connected to a similar use in relation to the dice throw and to chance that we saw in Deleuze's work on eternal return, in the previous chapter. Both these lines of thought on chance, games and time are developed to their fullest in the tenth series of *Logic of Sense*, 'of the ideal game'. One of the core lessons of that series is that any event communicates with all other events and does so in relation to the time of Aiôn, where chance and all events meet in 'one great Event' (LoSf, 81). This reflection on Aiôn and thought is important and we shall return to it in the next section. Because all events communicate, when we grow bigger, we also grow smaller, and grow wiser and duller, and older and younger, and faster and slower – all 'at the same time'. Another way of putting this is therefore that there cannot be a process of becoming in isolation. Each pure becoming is connected to all the others, not as negatives or opposites but as positive variations in themselves. This allows us to understand better the definition of pure becoming. A pure becoming is one defined in terms neither of oppositional relations, where fast is opposed to slow, nor negative ones, where tall is not short, for instance. Instead a pure becoming is a process all in its own right and it varies, not in relation to others according to logical or empirical relations, but in terms of intrinsic intensity, more or less slowness, for instance (which does not imply less or more speed). When a becoming varies in intensity in relation to an actual process – in the case of this series in relation to Lewis Carroll's Alice – all the other processes of becoming vary too. But this never takes place according to a pattern set outside their relation to Alice, including prior logical or empirical relations. Alice is growing bigger and smaller at the same time because she expresses an intensity of growing bigger and an intensity of growing smaller, as well as all other virtual ways of becoming.

We can now see why Deleuze says that a becoming avoids the present, because though it is in relation to a being in the present – Alice – its own variation and its relations to every other becoming take place in one play or blow outside that present. So when he says avoids, sidesteps or evades the present, *esquiver le présent*, it might be better to understand this as undoes the present. This is because the present is never separate from all other processes of pure becoming, nor before or after them. Instead, the present as an actual living present is undone in its expression of pure becoming, because each contraction of the past and of the future on to that living present is undone through the wider determination of all the pure ways of becoming. If the actual time is strictly one where the past and the future are dimensions of the present, for instance when all the past of a being is concentrated into the present wound drawing its life to an end, then this life is also nothing of the present at all through its participation in every pure becoming at its singular degrees of intensity. We are therefore being given an image of time where an event is only the present, with past and future as dimensions, and only past and future, where the present is a dimension of each one in different ways – and where each one also renders the other into a dimension of it. Once again, the manifold nature of time is important here and provides a response to the concern of priorities given to different times in Deleuze's philosophy. It is never that we have only the present, or only the past, or only the future, with other times as lesser dimensions. It is never that we have only Chronos or only Aiôn. Instead, all of these times coexist and together provide a complete view of time irreducible to any one of its elements or to an overall rule for their articulation.

Deleuze is well aware of the great challenges carried by this view of time. He uses a short study of Plato to address them, through an ambiguous take on the problem of pure becoming and on the role of common and good sense in Plato's philosophy, as means to tame and live with the impossible demands of pure becoming. The ambiguity is familiar with respect to Deleuze's take on Plato, since it is very close to many of the structures of Plato's thought, yet distant from their content. According to common sense, in its limited view of the actual Alice within categories of goods, we would say there cannot be all the processes at the same time in any actual being. *What are the goods for this species of being?* Each being has a right proportion of each becoming and these proportions must bend to the rules of logical oppositions. Too deep a cut and an animal bleeds to death. Too high a temperature and a rock can melt and transform irreversibly. Even if Alice can grow bigger and smaller at the same time, she must not do either too fast. Common sense, though, cannot determine action in relation to becoming in each circumstance. Its function is to define the possible relations between goods, but it cannot guide us in deciding on the right proportions for an individual being. Deleuze therefore defines good sense as the determination of the right becoming. *Which is it to be, in this case, bigger or smaller and at what speed?* The Platonic concepts of model and copy fit into this account of the senses, because the model is judged according to its participation in an original idea, whereas the copy is judged according to a ranking of copies. Which is the closest to the model? Which furthest away? A simulacrum on this account would be a copy gone bad, one furthest removed from its model due to a disfiguring pretence, or one that has forsaken its model.

However, the concept of a pure becoming and the accompanying erasure of the present demand a different account of the simulacrum. This version will be positive whereas the Platonic one is negative. Like the time of pure becoming, the simulacrum avoids the control of the ideal measure as set by common and good sense. It therefore also avoids the rule of judgement through the model and its copies: 'Pure becoming, that which has no limit, is the matter of the simulacrum insofar as it avoids the action of the Idea, insofar as it contests model *and* copy at the same time' (LoSf, 10). As we saw in the previous chapter, in *Difference and Repetition*, the simulacrum is the vehicle for pure difference in the operation of eternal return in series. It is the object = x that ran between pairs of series allowing them to be related but not through an identified object. In *Logic of Sense*, simulacra also have this freedom from identity, and this relation to series and to pure difference, now rendered as pure becoming. This allows Deleuze to show how simulacra undo and escape the rule of measure meted out according to common and good sense: 'Measured things are under Ideas; but under these selfsame things is there not still this mad element that subsists, that comes forth, out of reach of the order imposed by Ideas and received by things?' (LoSf, 10). Simulacra then complete pure becoming by providing a way of allowing difference into actual series, into Alice as actual series, but out of reach of common and good sense.[5] What, though, are we to make of Deleuze's description of the simulacrum as 'mad' element?

There is a becoming mad to this idea of the pure becoming and its conjuring of every becoming in the same play or time, because knowledge and direction are lost with order and comparison. There is no longer a right way and a wrong way, something Deleuze finds in Alice's characteristic unanswered questions: 'Which way? Which way?' How can we answer the question if we have to go all ways and if every way is also therefore the wrong way? Questions without determinate answer therefore generate a madness that comes out of the undoing of an order set according to an original model and its copies. We should be able to distinguish between right and wrong answers on the basis of which are the closest copies of the model, but once each answer is right *and* wrong this order is shattered. This results in an undoing of the present as the locus where we can act according to common and good sense, since any decision in the present is subject to the madness of not knowing the right answer. The madness of the simulacrum undoes the order that Deleuze analysed as a property of the Platonic circle in his study of the second synthesis of time. Instead of settled identified objects, each actual thing becomes series of simulacra, singular differences in their own right and in a manner always incomparable to any other. As in his work in *Difference and Repetition*, Deleuze does not flee the paradoxes that appear through the concepts of pure becoming, but rather sees them as generating another way of thinking about a problem. His name for the paradox generated with pure becoming is 'the paradox of infinite identity' (LoSf, 10). This must not be understood as the paradox of a thing that would have infinite identities, or that would be all things. Instead, it is the infinite identity or equivalence of all ways of becoming. They are all identical in response to the questions 'Which way?' and 'Which one?'

Why though is infinite identity a paradox? It is because when we allow for pure becoming we make every pure becoming identical in answer to the questions. Each one is the right answer. *It is this one and this one and this one.* They are all equivalent as right answers. However, this does not imply that the paradox indicates a dead end for thought. It means rather that the question posed was not the right one and that the understanding of becoming implied by exclusive questions is incorrect. *We should not be asking which one, but rather how much of each.* According to Deleuze's analysis, each becoming is pure, because when it is taken as having a given boundary it is also implied that it can be taken beyond the given limit; pure becoming is beyond any given limit. For instance, the injunction not to cut yourself too deeply implies that you can. Instead of a moral mode of thinking based on exclusive options indicated by the question 'Which one?', we therefore have a challenge of infinite mixtures and degrees, and the problem becomes 'Each one to which degree?' This is a practical question in relation to ways of becoming viewed as singular and incomparable, yet also as interlinked in relation to any action.

So my use of moral here is both unusual and open to question. It does not imply a morality as general set of laws prescribing well-determined general distinctions between different courses of action. Instead, it implies a singular moral and practical application of a series of principles drawn, in this case, from Deleuze's philosophy of time. 'Moral' here indicates an individual art of living, characterised by questions of dosage such as 'How much of this can this body bear?'[6] This version of moral, though, raises serious critical questions. Does it commit Deleuze's work on time to a form of individualism that breaks the collective or universal responsibility characteristic of moral laws? Does it commit Deleuze to a form of dangerous experimentation with limits, always pushing that bit further towards 'maximum intensity' across each 'pure becoming' as far as a body can take? Is Deleuze's philosophy of time, viewed morally, a rarefied version of selfish, thrill-seeking individualism? Has he abandoned Platonic common and good sense, only to sacrifice other beings to our own fatal quest for the maximisation of desire?

The first series of *Logic of Sense* closes with remarks that provide preliminary answers to these criticisms. Deleuze's philosophy of time does not lead to principles for a tight focus on individuals at the expense of a collective. On the contrary, the lesson to be drawn from his study of pure becoming is that the individual is necessarily lost in pure becoming. He states this through reflections on the proper name since, alongside time and the event, the philosophy of language is one of the main lines of enquiry in the book: 'The loss of proper name is the adventure repeated through all Alice's adventures. Because the proper name is guaranteed thanks to the permanence of a knowledge' (LoSf, 11). With the turn away from common sense and good sense, there is also a turn away from the permanence of knowledge in relation to moral principles. This is because each moral situation is singular and resistant to general knowledge. It is also because even in relation to the singular individual any knowledge is constantly erased through the work of pure becoming, that is, through novelty and change demanding new responses. The importance of the term adventure therefore lies in a strong sense of the term; it is an adventure into the completely new, into events demanding novel and unknown acts and choices. To turn to a selfsame individual, self or subject would be a fatal misunderstanding of Deleuze's philosophy. It is fatal in relation to the necessary perishing of the same:

> Because personal uncertainty is not a doubt external to what happens, but rather an objective structure of the event itself, insofar as it always goes in both ways at the same time, and tears the subject apart according to that double direction.
>
> (LoSf, 12)

The objective structure of the event is therefore not in an object in relation to a subject, but rather in the structure of series, simulacra and differential variations. This latter structure operates in the double direction of the concentration on the present and the dissipation into pure becoming. Both of these destroy the subject. The first does so as impersonal wound, a physical mark that leaves no space for personal identity. The second does so as impersonal purity, as a becoming beyond the limits that any subject or self could bear.

Torn apart in this way by every pure becoming, the challenge and test for subjects is not in how to resist the wounding and dismembering, but rather how best to ride with it. This is why it becomes a question of selecting all directions of becoming to participate in but at different intensities. The selection will undo even the subject making the choice, but this is not cause for despair at a negative contradiction, but rather a reflection of an ongoing transformation of the subject by the series and simulacra that constitute and alter it. The antidote to the focal wound is in the pure becoming. The antidote to the pure becoming is in the physical actual body, selecting through its wounds. In the version given in *Difference and*

Repetition, the underlying simulacra are the larval subjects and dark precursors underlying the subject, themselves expressions of the passive syntheses of time. So Deleuze's moral philosophy becomes a question of how to live with a necessary loss of identity. But does this not, once again, invite a criticism in terms of a perverse individualism: an individualism of one's transformations and process of becoming something different? *Are not some of the most selfish beings those embarked on their private adventures of discovery?* The answer is that this kind of individualism of personal adventure would be as big a misunderstanding of the reasons for the necessary transformation of the subject as any focus on maintaining personal identity. This is because pure becoming is expressed and participated in by all actual beings. An event and an adventure are then never private but necessarily collective. This can be seen in Deleuze's insistence on the communication of events. No event is monadic and isolated, and any selection in relation to an event and a becoming is a selection for all, not for one. But even if we accept this, does it mean we must select well for all in the absence of a general common sense and shared good sense?

TIME AND THE SURFACE BETWEEN DEPTHS AND HEIGHTS

A full idea of Deleuze's articulation of time as a balance of harmony and disharmony between deep physical incisions in the present and the purity of manifold ways of becoming requires an understanding of at least the following components of his wider philosophical system: his definition of singularity in relation to time and to intensity; his definition of the event as a process in time demanding a redoubling, replaying or counter-actualisation; and his account of the different ways Chronos and Aiôn determine one another. The concept of singularity allows for a further definition of individuation resistant to generality free of isolation and individualism.[7] Counter-actualisation explains how time is also a practice, how processes make time within time by replaying times differently – to the point of replaying all of time eternally. The relation of Chronos and Aiôn explains how the more familiar terms of past, present and future are translated into the philosophy of time from *Logic of Sense*. It therefore also allows for a better understanding of the connection of these terms to the three syntheses of time from *Difference and Repetition*. However, before studying each of these concepts in separate sections, I will look at how time is viewed in terms of wounds, infinitives and surface in the second series of *Logic of Sense*, in order to explain why time is essentially becoming rather than stasis.

As we saw in the second chapter on the first synthesis of time, Deleuze defines the present as a contraction. When translated into the language of *Logic of Sense* and into his study of the Stoics, the contraction becomes a physical wound. It is still a living present, but now one viewed in terms of the wounding mixture of bodies. It could seem that this is an important contrast between the two versions of Deleuze's philosophy of time, but in fact it is more of a change in emphasis based on a stronger focus on what happens in the difference between action and passion. More importantly, it is also a reflection on the stakes of the difference in levels and perspectives at work in whether we take a limited view of the contraction, or more precisely, whether we take a limited selection within it. The living present is always a matter of passion and action. It ensures the passage from the past to the future through passivity in relation to the past and the future, but also through an action, that is, a selection in relation to both. Any living present is made passively by the past and selects within it. The living present is also made by the future as a general possibility that the present selects particular paths within and, more significantly, through the eternal return of pure difference as novelty in the living present. Passivity is where we can begin to see time as a question of wounds and mixtures.

This passage from the second series of *The Logic of Sense* gives a version of how time is a form of bodily mixture and hence a form of wounding:

> The present is the only time of bodies and states of things. This is because the living present is the temporal extension accompanying the act, expressing and measuring the action of the agent, the passion of the patient. But a cosmic present embraces the entire universe, in relation to the unity of bodies among one another, in relation to the active and passive principles: only bodies exist in space, and only the present in time. There are no causes and effects among bodies: all bodies are causes, causes in relation to one another, one for another. The unity of causes among themselves is called Destiny, in the extension of a cosmic present.
>
> (LoSf, 13)

This difficult and dense description of the present is given as a commentary on a Stoic distinction between two types of thing: bodies and infinitives. Infinitives are Deleuze's way of presenting pure becoming in language.[8] Each pure becoming is rendered by an infinitive such as 'to grow'. Each subject grows in this or that way

Time in Logic of Sense

and at this or that intensity, but all participate in the infinitive 'to grow' they give their individual expression to.

The two types of thing rendered in the passage correspond to two types of being: existence and subsistence. Things exist in time. Infinitives subsist in time. However, this distinction must on no account be confused with a distinction between real and unreal beings and Deleuze's philosophy should on no account be confused with an idealism of subsistent infinitives or a materialism of things. On the contrary, reality is both existence and subsistence and if things are taken as merely existent or merely subsistent, they are incomplete realities. Deleuze's philosophy is viewed better as a speculative dialectics of things and infinitives or pure becoming, than as resting ultimately on either ideas or matter. Existence and subsistence complete one another. This is one of the points that Deleuze is making with this description of the present. The point also therefore indicates another possible misunderstanding. It could be understood that Chronos and Aiôn or the present, past and future are being separated according to things and infinitives, where the present is only of actual things and past and future are only of infinitives. It is almost exactly the opposite: *actual things are only of the present and infinitives are only of the future, but the present and the future are for each other such that things still take place in relation to the future and infinitives still require expression in the present of things*. This is shown for things, in a first step, by the cosmic present embracing the entire universe and by the definition of Destiny.

That the present embraces the entire universe corresponds to the way in which the living present contracts the whole universe in the first synthesis of time in *Difference and Repetition*. However, the passage from *Logic of Sense* is extremely difficult and laden with translation problems due to Deleuze's use of the expression '*à la mesure de*' often translated as 'commensurate with' and given as 'to the degree that' in the current translation (LoSe, 4). Both these translations introduce a conception of variation in the relation to the entire universe and even the idea that it could be the case that some living presents do not contract the entirety, if the degree is zero, for instance. I do not think that variation in degree is Deleuze's point, nor a condition for the contraction of an entirety. Instead, unlike the current translation which puts the degree subordinate clause first, my view is that Deleuze is stating the contraction of the entirety then stating that it varies in terms of its qualities as a contraction of the entirety. In other words, all living presents contract the entire universe, but each one does so differently in relation to different processes of unification according to principles of activity and passivity. Within a limited selection necessary for the concept of acts, agents, passions and patients, the present is a site for mixtures of bodies defined by acts and passions. Each body is defined by its acts, by its causes. When these are limited by the acts of other bodies, in a mixture of bodies, the result is a passion and each body is therefore also a mixture of action and passion, an agent and a patient. The living present is therefore necessarily a time of wounds brought about by the clash of causes among each other. This clash has the entire universe as horizon, a present determined by all causes, wherein the destiny of any thing can be read as its place in all mixtures.

However, this present of bodies, mixtures and destiny is incomplete and must be viewed with a necessary complement of the effects of the mixtures outside the time of the living and cosmic present. This different time is defined by Deleuze as the time of incorporeal effects, that is, of the changes in the intensities of relations of infinitives. The key to this difficult argument can be detected in the novel and strange use of the ideas of cause and effect, and existence and subsistence. Modern philosophy views these terms as linked pairs: a cause for an effect and ideal subsistence as opposed to full real or material existence. Deleuze separates them and in so doing raises two conundrums. How can there be causes without effects and effects without causes? How can there be subsistence that is not a deprivation of existence and an existence that is an addition to subsistence, for instance, through an idea being realised? The answer lies in the extreme novelty and originality of his philosophy of time and of his version of reality. There are only causes in the living present because each existent is only defined by its synthesis of an actual series. To cause is to be a synthesis and the synthesis is not an effect of an external cause. The many bodily syntheses cross over one another and limit one another, but they do not cause one another. The living present is a competition of causes, bringing them together as agents and patients: the cause of the weed killer, limited and wounded by the resistant strain of grass, itself cause of a synthesis limited by the cause of drying land, itself cause of lines and patterns across a landscape, itself limited by the wind as cause, itself limited by the contours of the land. The destiny of the weed and the destiny of the eroded hillock all read into a cosmic mixture of causes in the extended living present. To exist is to exist as a cause and as a passion, with a set destiny among other causes.

Time in Logic of Sense

But existence is not all there is, because the sense or value of the bodily mixtures is not in their existence as causes, but in their effects on infinitives, in the value or intensities given to each mixture through its effect on the changes in intensities of each and every infinitive. This explains why Deleuze distinguishes existence and subsistence: the former is reserved for actual things and their existence as causes; the latter is reserved for infinitives or ways of pure becoming and their subsistence as effects. As a field grows brown and barren there can sometimes be a diminishment of the intensity of 'to grow', but at another time, that is, in another living present, there can be an increase. The measure of these intensities is not in the actual field, but rather in the effects as played out in other expressions of the infinitives it has effects on. The effect on sense and on the infinitives is not the limitation of causes among themselves in the present, but rather a forward and backward effect of the causal events on sense. So any event is now divided into two processes in time: one of bodily mixtures and another of changes in the intensities associated with infinitives.

To return to the philosophy of time, bodies only operate as causes in the living present. Infinitives only operate as effects in the past and in the future. Yet neither bodies nor infinitives are complete without the other:

> So much so that time must be grasped twice, in two complementary ways, exclusive of one another: wholly as living present in bodies that act and submit, but also wholly as an instance infinitely divisible in past-future, in the incorporeal effects resulting from bodies, from their actions and passions. Only the present exists in time and draws together, absorbs the past and future; but the past and future alone subsist in time, and infinitely divide each present. Not three successive dimensions, but two simultaneous readings of time.
>
> (LoSf, 14)

To unpick this important summary of Deleuze's philosophy of time I will first pose a number of sceptical and critical questions, in order to see if it is possible to give an interpretation answering them:

1. How can there be two complementary ways of grasping time that are both whole?
2. How can there be a time that is wholly present and another that is wholly past and future?
3. What are incorporeal effects and how do they relate to corporeal causes?

4. What does it mean to have two simultaneous readings of time that are also exclusive of one another?
5. How does an account of two simultaneous readings of time that are not of three successive dimensions relate to the concept of dimensions of time and three syntheses of time from *Difference and Repetition*? Did Deleuze change his mind between the two books, or did he maintain two inconsistent philosophies of time?

Chronos, the time of the living present, and Aiôn, the time of infinitives, are whole yet complementary to one another because they are whole or entire on their own grounds under one aspect, yet completed by the other time, under another aspect. As a synthesis of time that concentrates it on the present alone, the living present is whole since it is a synthesis of all of the actual past and all of the actual future in the present. In this synthesis, past and future are dimensions of the present and it contracts them in their entirety (according to the syntheses of time given in *Difference and Repetition*). However, this synthesis is not itself complete until it is determined as to how it itself passes away and how it becomes into the future. The present is complete as a synthesis of its own dimensions but it is incomplete in relation to other syntheses that determine and explain some of its features: its passing away and its internal novelty. This determination is only achieved when the effects of a living present are taken into account and these effects require a move into Aiôn. Each living present needs to be determined in terms of how it will pass away, rather than how the past is a dimension of its living contraction. Furthermore, each living present needs to be determined as to how it participates in a new future, rather than how the future is a particular dimension of that living present. The time of the past and the time of the future, of the always has been and the always to come, are also whole. As pure past or as every pure becoming at different intensities that will have been, the past is entire and lacks nothing. As every pure becoming that will be, the future is whole. However, this past and this future require an actual expression to be complete, that is, to be determined as this past and this future, rather than the pure past and the future of pure difference. So though Chronos and Aiôn are both complete, the former requires the latter to move outside itself into the past and into the future, whereas the latter requires the former to move from a pure and chaotic potential to a fully determined set of relations.

So when you apply a tourniquet to your arm to stop a deadly flow of blood all of your past and all of the past of all the actual things around you are concentrated in your struggle. This is also true of all the particular futures held in your existence, the possible things that can and cannot happen now that you are there wounded – not only those things for you, but those things for all the actual things around you. This is because all of the past and all of the future are taken in relation to that wound, concentrated in it. The present wound is the whole of time under the perspective of that wound. Other wounds and other beings will also be the whole of time under their singular contractions of the past and of the future. The hand that held the knife as it cut into your flesh is its own living present, contracting the past and the future differently on a different present to the time of your fatal wound. Yet this is only one side of the wound. It is the way it concentrates all of the actual past and all of the actual future (the future as possibility) under its perspective. In addition, the wound has an effect on the pure past as it passes away. It has an effect on the future in this passing away and in the novelty of the event of the wound. Your actual struggle might strengthen the relations of the infinitives 'to struggle' and 'to flourish' for other wounds in their presents. Why might? Because virtual effects are only completed when they are expressed anew or counter-actualised – until they are, they remain as potential. These effects complement the first actual synthesis with the way in which any present alters relations of infinitives or pure becoming. We can, with some risks, describe these as relations of significance and value. Any present event is then an actual cause and a virtual effect. It is a synthesis of all actual events, but it is also an effect on the intensities of the pure ways of becoming that can be expressed in other actual events, in their passing away and their own novelty.

But was not the thrust of that hand the cause of your death and the death its effect? Not from the point of view of each living present, where the seeping wound is cause of its contractions but not those around the murderous hand, and where the aching hand is cause only of its contractions, including your flesh as it splits under the pressure and edge of the knife, but not its own living present as cause and synthesis. There are only causes in the present of Chronos because each living present contracts others as cause, but not as their own living presents. We therefore have an actual world of multiple perspectives where each perspective is a cause in the way it is a synthesis of its world, but not a cause in the way other perspectives are synthesised. More precisely, though in a somewhat baroque turn of phrase, we have many worlds in multiple disjunctive syntheses of series.

The concept of disjunctive synthesis is justified because it explains how each living present synthesises the entirety of actual and virtual events, but does so in its singular way that creates a disjunction with other series, such that it cannot belong to them or be reduced to them. There is a synthesis, but it is not one of resolution or reduction, but rather one of addition and radical novelty. So all series include one another, but a living present never includes another as another living present, because a living present is a synthesis and contraction and not a moment in the contraction of another present. The series are disjunctive, not only at the point of the living present, where the new disrupts the pattern of the past, but through every member of the series altered by simulacra, and hence pure differences, as they travel back and forth through the whole series. Disjunction is then not a division, found for instance in the decision between two known possibilities, but rather a transformation leaving nothing unchanged and nothing within the secure grasp of prior knowledge and later representation.

Incorporeal or virtual effects explain and determine the singularities of each disjunctive synthesis. They determine why each living present is singular and hence a disjunction. A living present is different from all others through the way it passes into the pure past and expresses pure becoming in its novelty. Pure difference is experienced in the living present of Chronos as singularity, but it does not find its reason there. Instead, the individuation occurring with each synthesis is determined not by chains of cause and effect in actual series, but rather within the changing intensities of pure becoming or infinitives in the virtual or subsistent realm of Aiôn. Each pure present and its take on the whole of actual time are therefore accompanied by a result in Aiôn, its effects in the intensities of the pure realm. So though different present syntheses fail to communicate as actual, each determining Chronos in its singular manner and never taking the others as living presents, or causing them to change, those syntheses communicate through their transformations of Aiôn.

All events draw on and acquire intensity through a shared realm of sense, of infinitives and pure becoming. This realm is never actual and the communication is asymmetrical, that is, a communication through shared effects and transformed intensities, but not through the reversible symmetry of cause and effect. *The film of Deleuze's reality can never be run backwards. It moves forward through the work of the new and at the same time moves backwards as a wave of changing intensity through the pure past.* Neither the pure differences returning in the future, nor those in the pure past absorbing each passing present remain the same, to be replayed and recaptured as the same. Instead, the only communication with the past and with the future afforded to actual living presents – to us – is through our attempts to replay differently, to counter-actualise, the ways in which other living presents touched on and expressed pure becoming. We only communicate by becoming different. Every life is always an experiment with pure difference through novel creation.

Instead of a monotone or homogeneous view of time and reality Deleuze gives us a multifaceted and multiple idea of time, where the real divides according to two times.[9] These times cannot be reduced to one another and are self-sufficient on their own terms yet complete one another in order to give the full version of reality. This explains why Deleuze can claim the two times as exclusive of one another. The future in Chronos is not the same as the future in Aiôn, for example. It also explains why he insists on more than one time despite this exclusivity and the self-sufficiency of each time. Whole and complete do not mean the same thing. The present, with its dimensions of past and future, is whole in Chronos, but incomplete until it is considered with Aiôn, which is also whole and free of the present, except as dimensions of the past and of the future. But is not the concept of dimension foreign to *Logic of Sense*? Has the interpretation given here not imported terms and views from *Difference and Repetition* to give an illegitimate reading of its sister book, where illegitimate means untrue to the exact letter of a book?

I have little patience with the idea of a legitimate reading based solely on the boundaries set by the covers and letter of a book. The point of an interpretation is to explain and perhaps enhance, to connect and differentiate, to exemplify and add voices, to chime with and offer counterpoint, to develop and to unpick, to analyse and give reasons, to criticise and, with luck, to expand. Consistent with Deleuze's concept of counter-actualisation and with his theory of time as requiring novel difference for its syntheses, these goals of interpretation are falsely hindered by arbitrary limits – arbitrary because they find little justification in works based on connection and contrast, such as Deleuze's three great books (the third, *Expressionism in Philosophy: Spinoza* has been underplayed here, except in the study of the use of the concept of adequacy in *Difference and Repetition*). Yet within the aims of a more broad and open reading there are still potential difficulties in mixing terms from two books, as I have done. Some of these appear convincing and stark because they are based on large conceptual contrasts: Chronos and Aiôn versus the three dimensions of time; the lack of a central role for Nietzschean eternal return in *Logic of Sense*; the stronger human and moral focus of that book when compared with *Difference and Repetition*; arguably, the greater role given to psychoanalysis in *Logic of Sense*.

The first of these contrasts is instructive for the rest. In *Logic of Sense*, when Deleuze denies that there are three successive dimensions of time, this is consistent with his conception of dimensions in *Difference and Repetition*. The denial is about *successive* dimensions, which must be replaced by *synthetic* dimensions, for instance where the present contracts the past and the future as dimensions. The insistence on two times is not inconsistent with three syntheses because the processes are what matters and not the times as such. In this sense, Chronos and Aiôn are more likely to mislead interpreters than the idea of three syntheses because they still allow for the idea of two times as separate containers or states. This is a misunderstanding, though, because Chronos and Aiôn are connected through determinations which mean that they cannot be separated. Furthermore, they are connected not in one or two simple ways, but as multiple processes on each side of time and between them. The test of consistency for Deleuze's books, concepts and ideas is almost never about categories and numbers, but rather about process and system. How do events work here in comparison with there? In answer to this kind of question, there is an extraordinary consistency between his two books and two ways of describing time: contraction in the present; asymmetrical processes between actual and virtual realms; changes in intensity in a realm of pure becoming, itself divided into past and future processes; and, most important of all, connection of all of these reciprocal determinations in a complete account of reality.

FROM PRINCIPLES TO ACTS

The fullest series on time in *Logic of Sense* is the twenty-third series, 'of the Aiôn'. I will now study this series with a view to answering the remaining critical questions about Deleuze's philosophy of time as presented in that book, with the additional aim of explaining how the philosophy translated from principles about time to principles about action. There is no denying the difficulty of Deleuze's account of Chronos and Aiôn. Demonstrations of how it relates to work in

Difference and Repetition and of how it provides responses to critical questions about consistency provide some clarity, but further explanation is necessary if we are to understand how the philosophy of time relates to action and to selection. The point now will be to give a somewhat simplified version of the philosophy, through a set of theoretical principles, and a practical context, through a set of practical principles.

The 'of the Aiôn' series is divided into a study of Chronos, followed by a study of Aiôn. This division is a trap, however, because one of Deleuze's main arguments is that two opposed readings of time are required for a complete view of time. Furthermore, on closer inspection, it becomes clear that although the two readings are indeed opposed (and perhaps it is better to say heterogeneous, here) the two readings are also related and indeed dependent upon one another. So the use of the term complete, in this instance, is not in the sense of complete through the union of two separate but consistent parts, but rather complete through the full connection of many interlinked yet fundamentally irreducible processes across two readings of time. It could be objected at this point that my interpretation of this philosophy of time is obviously wrong in the light of statements made by Deleuze, such as in the expression 'Aiôn against Chronos' (LoSf, 191). The answer to this objection rests partly on the moral and practical tenor of Deleuze's work on time in *Logic of Sense*. He is devising a version of his account of time explaining how a rift in time impacts on bodies, ideas and values, thereby also impacting on how we should react to events. This rift introduces a tension and set of oppositions in any situation and in any response to it, such that his philosophy of time is not only conceptually difficult but intrinsically demanding in its implications for the condition actors find themselves in, due to the nature of time or, in reality, times.

My concern in charting this series of principles about time is not to focus on the most fundamental formal aspects of Deleuze's philosophy of time as set out in *Logic of Sense*. Most of these, such as the claim about only the present existing in actuality, have been covered already here and at length. Instead, I want to dwell on many of the implications of these formal characteristics, since the twenty-third series presents these better than anywhere else in Deleuze's works. Here are the philosophical principles, beginning with the ones drawn out in relation to Chronos, then moving on to those given in the sections on Aiôn:

1. Each time we define an actual past and present in relation to a bounded present, we can devise a more extended present including that past and future within its extent and duration (LoSf, 190).
2. Presents are relative to one another according to an order of inclusion, but where the operation of inclusion is also a transformation. Each present has a duration of its own and another duration when it is included in another present (LoSf, 190).
3. Presents are lived differently according to the limits set for them. Each relation of inclusion is a complication of more limited presents, where complication is defined as inclusion without equivalence between an element and its included state (LoSf, 190).
4. The present is the time of mixtures of bodies implied by the relations of inclusion. To be included in a wider present is to be passive in relation to its actions of inclusion of other more limited presents. Every present is a limitation and a resetting of all that it includes (LoSf, 190–1).
5. The present is fundamentally troubled because it is a search for the right measure, in response to mixtures of bodies and struggles between action and passion, where this measure is always lacking (LoSf, 191).
6. Every present is divided in two directions. On the one hand, it is pulled towards a good measure. On the other hand, it is directed to pure ways of becoming, as the past and the future, as expressed in present. The present is a locus of division and rift, not only in its mixtures, but also in its directedness towards the pure past and future (LoSf, 192).
7. When thought in relation to pure becoming, every present duration is divided by an instant drawing it into the past and into the future, such that each present is not only a search for the good measure, but also an attraction towards an act that is adequate to the pure past and to the test of eternal return (LoSf, 192).
8. There are two forms of adequacy in time: the right mixture for a given present duration and the adequate selection for all of the past and all of the future (LoSf, 193).
9. The instant is incorporeal. It is not of the mix of durations but of the pure and empty forms of time; its task is to pervert the whole of the past and the whole of the future instantaneously and eternally (LoSf, 194).
10. Time is an event in bodies and an event in pure becoming; in

Time in Logic of Sense

bodies it is a novel event of mixtures and limits; in pure becoming it is an event drawing together all of the past and all of the future, as well as every present mixture (LoSf, 194–5).
11. The coming together of bodies and incorporeal becoming takes place through singularities and through shared intensities (LoSf, 195).
12. Every act takes place according to two readings of time that cannot be reconciled yet that must coexist according to their own forms of adequacy (LoSf, 196).

Here is a set of principles guiding action in relation to the principles taken from Deleuze's work on Chronos and Aiôn. These principles shape an underlying moral problem and its determining questions as they emerge out of Deleuze's philosophy of time:

1. Any act must take account of the ongoing durations and boundaries it rests upon and also those that include it in their own durations. *How have you included others in your present? Whom have you excluded?*
2. When we act according to our durations we transform those we include within them, as we are included in those of others. *How are you transforming? How are you transforming others around your present wounds?*
3. There is no neutral relation of inclusion between durations. *Whom are you hurting? Who is hurting you?*
4. Any act is active in relation to some durations and passive in relation to others. *How are you making time? What times are making you?*
5. We have to search for acts that bring an adequate measure to the conflicts between different durations even though no such adequacy can be achieved. *How can we achieve the widest reciprocal inclusion adequate to our shared wounds? How will it fail?*
6. We have to search for acts that move beyond the durations they rest upon and that transform relations of inclusion beyond any adequate measure. *How can we move beyond our presents? How can we transform our relations beyond any possession?*
7. An act moving beyond its durations must seek an adequacy with all potential ways of becoming where adequacy means to affirm only those that are pure. *How can we be worthy of all the intensities and all the relations of becoming expressed through us?*
8. An act must seek both an adequate measure for actual durations and their mixtures, and an adequacy to all of the past and

all of the future. *How will this act be worthy of our present's wounds as well as the eternal becoming expressed in them?*

9. An act in relation to pure becoming takes place as an instant rather than a duration, it is a novel creation that then travels through the whole of time. *How is your present act eternal, changing all of time, all of the past and all of the future instantaneously?*
10. Acts are event-like not only in bodies, but also in thoughts in their perverse relation to all past and future events. *How is this thought eternal, changing all of time, all of the past and all of the future instantaneously?*
11. The way into pure becoming is through the detection of singularities in relation to intensities. *Where are your singularities, your dark precursors, your larval selves headed to?*
12. Every act has to be true to a present duration and to its inclusions as well as to an eternal time and to the way an instant determined by the act alters the whole of time. *How will your act be worthy of its inclusions, yet also worthy of its eternal legacy? How will you be worthy of your double destiny as a passing self and an eternal return?*

Such principles and questions are constructed into a difficult multiplicity. They know of no final answers. They do not submit to paradigms or analogies. They are the moral problem not as mental puzzle or political foil, but as challenge to always live differently in accord with events but never with final judgement over them.

7

Conclusion: the place of film in Deleuze's philosophy of time

The most extensive application of Deleuze's philosophy of time appeared some fifteen or so years after the philosophy was developed in *Difference and Repetition* and *Logic of Sense*. In the second book of his works on cinema, *Cinema 2: The Time-Image*, Deleuze returned to his work on Bergson and on the pure past in order to show how film can convey the multiplicity of times, their relationships and processes. This study turns on a concept that has antecedents in *Difference and Repetition* but in a lesser role: the crystal, or crystal of time. *Cinema 2* also has one of the most succinct summaries of the many dimensions of time, as studied in depth here. This statement is made through a reference to Augustine that does not appear in that form in the earlier works by Deleuze. This is a characteristically self-effacing gesture by Deleuze, since everything he draws from Augustine is in fact treated in a deep and original manner in the three syntheses of time from *Difference and Repetition*. So why have there been no references to Deleuze's works on cinema in my study of his philosophy of time? Should not those works be used for a full understanding of that philosophy not only as a further source, but as the canonical application of the philosophy? Why bypass one of the most prolific areas of Deleuze interpretation and one of the places where his influence has grown fastest and with the strongest effects not only on theory, but also in practice?

The reservations I have with using the cinema books here are not generated by their value as works on film.[1] On the contrary, these are indeed rich and extraordinarily subtle conjunctions of philosophy and film. However, as explanations of the full range of Deleuze's work on time they fall short for a number of reasons. I will look at three of these here. First, the focus on the image is problematic. This is not due to the concept itself, since it has a careful elaboration notably in chapter III of *Difference and Repetition* and in Deleuze's works on Bergson. It is caused by the conflation of the philosophical use of image, where it stands for a necessary yet risk-laden restriction of intensity and ideas, and the cinematic image, where, however much we seek to expand it outside the screen, to the brain, to senses, to perception, to thought, we still retain screen images as the prompt, support and central reference for these wider processes. This means that representation retains some of its force over the formal metaphysics developed in the earlier works on the philosophy of time. We find this in the language of *Cinema 1* and *Cinema 2*, where Deleuze's reading is dominated by a language of reference and representation: 'What the lover said', 'What the prince heard too late', 'What Visconti did not show us', 'The too late that gave rhythm to the images', 'The revelation of the musician in', 'The most beautiful scenes in' (Deleuze, 1985: 126–7). These phrases come from one paragraph of Cinema 2, from chapter 4 on the crystals of time, yet they are indications of a general approach for both cinema books. They tie processes of thought, developed with great insight by Deleuze, to images and scenes from Visconti's films. My concern is not strictly with this operation of reference; it is with its effect on the philosophical concepts, since no matter how hard we try to avoid associating a primary philosophical image, such as the image of thought from *Difference and Repetition*, to the film image, we still retain the representational frame as our first point of understanding and explanation. It is an inherent weakness of thinking through film, as opposed to the more speculative and pure work from *Difference and Repetition* and *Logic of Sense*.

It could be objected at this point that works such as *Logic of Sense* have many literary references. Do these not tie Deleuze's philosophy to representation as strongly as the later film works? The answer is that the books on cinema are much more uniform and traditional in their forms of reference. Deleuze's style in the cinema books is frequently restricted and descriptive, in the sense of alternating short sections of theory and much longer sets of references for it. This is a long way from his style in *Logic of Sense*, where different registers are interwoven, such that passages of reference and of speculative elaboration cannot be separated. The original conceptual construction takes place with the literary works, for instance, through Lewis Carroll. This weave is far less apparent in the cinema works and the bridging between registers leads to a frame of indication, of theory and target, such that the idea of image is much harder to free from an inner representation and memory of images and scenes from the films. It is this separation, brought about by a work on one art form approached through a great number of its masterworks, that ties Deleuze's thought and style to an 'aboutness' and a categorisation that we do not find anywhere else in his works. In some ways this is an advantage of the works on cinema, since it has allowed them to be taken up as templates for film analysis in a much smoother manner than, for instance, Deleuze's work on the concept of genesis in relation to evolution, or his concept of destiny in relation to literature. The smoothness is also a potential hindrance though, because the demand to create original works is tempered by the promise held by ever wider applications of Deleuze's concepts. Perhaps this also explains why many of the best commentators on Deleuze's work on cinema reconstruct the underlying philosophy on the basis of other more purely philosophical works.[2] It also explains why the time-image and movement-image distinction from the cinema books has come in for some strong criticism.[3] In contrast, the direction and separation of language and reference are much less apparent in Deleuze's book on painting, *The Logic of Sensation*, or in his works on literature such as *Essays Critical and Clinical*. There, concepts and artwork grow inwards and explode outwards together, in a style with much more rhythm, texture, complexity of pace and linguistic invention.

None of this of course diminishes the value of Deleuze's works on film.[4] Nor is it meant to. It is rather an explanation of why the film works add little and in fact might take away from his philosophy of time in its most consistent and extensive form. They are the closest Deleuze came to a theoretical tracing in art rather than a co-creation. The problem with such operations of marking out and sketching of outlines, of taxonomy (Deleuze, 1983: 7) is that they tend to restrict both the theoretical support and the traced aesthetic object. This comes out very strongly in the contrasting roles played by Bergson's work on memory and the pure past in *Difference and Repetition* and in *Cinema 2*. This is the second reason why the cinema books fall short of a satisfactory rendition of Deleuze's philosophy of time. As we saw in Chapter 2, Bergson's work is a point of departure, a prompt and a partial moment in Deleuze's work on the second synthesis of time. It is only an element of his work on the second synthesis of time, one that requires an alignment with Hume, the creation of a set of original principles about the past, and a critical departure from Bergson's own critique of representational and geometrical accounts of consciousness. There is also a considerable methodological contrast, most notably around transcendental deductions and the manifold dimensions of time, including Nietzsche's eternal return.[5] This disjunction between the philosophical and film work comes out very strongly in the sections on Bergson around the crystal of time, where the earlier distinctions between foundation and founding of time are replaced by the idea of 'the most fundamental operation of time', where time 'becomes double at each instant as present and past' (Deleuze, 1985: 109). As we saw through this study, that notion of the instant is not consistent with Deleuze's account of time in any of its dimensions. These sections of the cinema books are then more Deleuze's demonstration of the fruitfulness of Bergson's account of time, rather than his own. This view is reinforced when Deleuze speaks of 'seeing' time in the image-crystal: an expression that runs counter to the approach of *Difference and Repetition*, which treats experience and sensibility in a much more controlled manner, extending it thanks to operations of thought rather than special experiences.

It is only in the next paragraph of *Cinema 2* that Deleuze returns to his earlier philosophy of time in a summary of the principles he developed around Bergson. But the two sets of principles are different and the ones in *Cinema 2* are damaging to the plurality of times in *Difference and Repetition*. In the earlier book, the different dimensions of time are consistent without being reducible to one another. This is also true of *Logic of Sense*, for example, where a concept of the instant is used, but as closer to the process of differentiation associated with the simulacrum as it synthesises all of the past and all of the present by running through them all at once. In the cinema books, though, the concept construction is much looser and open to metaphorical, representational and analogical interpretations. So, there, the concept of instant is allowed in relation to the present, something inconsistent not only with the idea of the living present, but also with the concept of duration. Similarly, Deleuze speaks of time as interior, in the sense of that which we are 'interior to'. This spatial metaphor contradicts the much more complex but also more satisfactory ideas of making time from the earlier works, where the present is made to pass by the pure past, but where the present is never internal to the past in any sense of containment. It is rather in its grip and transformed by it. These interpretative stresses reach their highest point in the work on cinema where we have to reflect on the tension between Deleuze's use of pure in relation to the virtual in the work on cinema and

Conclusion

Bergson and his earlier work on the pure past. The virtual or pure past in *Difference and Repetition* and *Logic of Sense* is strictly levels of relations of pure becoming at varying intensities, or relations of infinitives and surface intensities. But in the cinema books this is equivocated by reference to the crystal and to the image. This equivocation is inherited from Bergson, where the concept of pure memory is opposed to conscious memory yet remains a memory in its form. Thus, instead of levels of the pure past, the cinema book speaks of regions we can travel to in film (for example, in Fellini). The regions all coexist, but the damage is already done with the notion of 'jumping into the past', an analogy taken from film but catastrophic for Deleuze's philosophy of time, since it confuses the present contraction of the past with the synthesis of the pure past that makes that present pass (Deleuze, 1985: 130).

The third reason to be wary of the cinema books as a principal source for Deleuze's philosophy of time is that, even when there appears to be a strong correlation between the work on cinema and the earlier philosophy of time, the later work encourages a fusion of the dimensions of time in images and events. This is a departure from the account of time in terms of sides and dimensions from *Difference and Repetition* and *Logic of Sense*, where the desire to view all the dimensions in one is always resisted. So instead of the network of asymmetrical processes from the earlier work, we encounter conceptions of simultaneity raising numerous critical points. Here is the closest we come to a summary of Deleuze's philosophy of time, as expressed in *Cinema 2*: 'According to Augustine's beautiful formula, *there is a present of the future, a present of the present, a present of the past,* all implicated in the event, rolled into the event, thus simultaneous, inexplicable' (Deleuze, 1985: 132). The problem is that this is an impressionistic and loose condensing of the earlier philosophy of time. It raises a puzzle rather than a productive paradox by using a concept of time, simultaneity, to draw together the three presents.[6] We have to reflect on the time implied by that simultaneity, whereas the three syntheses of time, and the asymmetrical reciprocal determination of Chronos and Aiôn, strictly avoided any notion of their simultaneity and hence of a further time in which they happen, however paradoxical that time would then prove to be. The appeal to simultaneity and to images then erases the asymmetries and orders of dimensions of the present as contraction, the present as contracted and made to pass (as dimension of the past), and the present as eternally returning as difference and never returning as the same (as dimension of the future). They are replaced by a much more simple and general relation of attribution (the present *of*) which itself erases the irreducibly different relations from *Difference and Repetition*.

Finally, the one great Event from *Difference and Repetition* and *Logic of Sense* is in danger of being reduced to a set of states of events:

> An accident will happen, happens, has happened; but it is also at the same time that it will take place, has taken place, is taking place; so much so that, having to take place, it did not take place, and, taking place, will not take place . . ., etc.
>
> (Deleuze, 1985: 132)

As Deleuze will then say, this raises a paradox, but it does so in a much thinner context that draws it to concepts of the simultaneity of signs and of images fatally attached to the more commonsense and experiential modes of exemplification and explanation we would encounter in film: 'Two people come to know each other, but already knew each other and do not yet know each other' (Deleuze, 1985: 132). This draws us into a line of thought governed by the question 'How is this possible?' and the answer 'As revealed in this or that film in this or that manner'. Against this revelation in experience and through the image (however much it is allowed to go beyond representation) Deleuze's philosophy of time creates with a different underlying problem: how can time be multiple in itself and generate multiplicities while resisting any reduction to a space–time continuum? The original and brilliant answer is that time must be a multiplicity of processes, where times are dimensions of one another according to asymmetrical syntheses. This is a time of resistance to settlement and to wholeness. It is a time forever inviting new, dizzying and ephemeral constructions: 'Thus ends the history of time: it undoes its physical or natural circles as too well-centred; it then forms a straight line, but one driven by its longueurs to reform an eternally decentred circle' (DRf, 152–3).

Endnotes

1 INTRODUCTION

1. For a starting point to work on Deleuze and Whitehead on time, see Williams, 2009.
2. See Michael Armstrong's *A Handbook of Human Resource Management Practice* for an overview of the types of objectives (Armstrong, 2006: 505–6) and in particular his account of SMART objectives in *Managing People: A Practical Guide for Line Managers* (Stretching/Specific, Measurable, Agreed, Relevant/Realistic, Time-related), as these make the connection to time most explicit (Armstrong, 1998: 151).

2 THE FIRST SYNTHESIS OF TIME

1. Deleuze's argument is close here to Hume's work on the problem of induction. We cannot conclude from past experience that the future will be the same as the past, even in terms of the association of causes and effects 'Reason can never show us the connexion of one object with another, tho' aided by experience, and the observation of their constant conjunction in all past instances' (Hume, 2009: 64). Deleuze's interest, in *Difference and Repetition* and earlier in *Empiricism and Subjectivity*, lies in the next step of Hume's argument and in the situation of the connection in the mind through constant conjunction (see Deleuze, 1953: 4–5). For an extended study of Deleuze's relation to Hume see Bell, 2009. Bell gives an insightful explanation of the similarities and differences between Deleuze's reading of Hume on belief and the mind and contemporary debates (Bell, 2009: 26–9). For other helpful discussions of Deleuze in relation to Hume, see Panagia, 2006; Roffe, 2009.
2. Deleuze returns to Leibniz's law and to his principles at much greater

depth in *The Fold: Leibniz and the Baroque*. See in particular his discussion of the mannerism of substances in Leibniz (Deleuze, 1988: 77). For a deep discussion of Deleuze's work on Leibniz's principles and, in particular, the principle of the indiscernibility of identicals, see Smith, 2009. Smith's reading has great value in relation to this discussion of time. This is because he shows how Deleuze's shift from identical substances to individuals is accompanied by a shift from concepts to Ideas (Smith, 2009: 54). This is important in regard to the relation between the first synthesis of time (an actual synthesis in the present) and the second (an ideal synthesis in the pure past). Juliette Simont also has an interesting discussion of Leibniz's law in relation to Kant and to the concept of intensity (Simont, 2003: 33).
3. Deleuze, following Hume, assigns this work of synthesis to the imagination: 'The imagination has the command over all its ideas, and can join, and mix, and vary them in all the ways possible' (Hume, 2009: 68). However, it is interesting to note that Hume's treatment is in the context of belief, a term used infrequently by Deleuze, perhaps because the reference to intensity of ideas associated with belief in Hume is treated very differently by Deleuze. Bell has an excellent response to this problem situating reason within the imagination for Hume, where other commentators separate the two faculties of the mind: 'Stated in our terms, the beliefs generated by reason and the imagination are each, as identifiable beliefs, inseparable from the historical ontology of the imagination' (Bell, 2009: 41).
4. For an excellent explanation of contraction in the first synthesis of time, but set in relation to organic life, in contrast to contraction for any being as presented in my arguments, see Ansell Pearson, 2002: 'The presentation of time he is developing is by no means restricted to human time – the contraction of habits through an originary contemplation is a feature of organic life in general' (Ansell Pearson, 2002: 186). In my view, this extension must go beyond organic life.
5. Perhaps the best known account of the problem of the present is Augustine's in Book XI of the *Confessions*, 'Time and eternity'. Augustine's treatment relies on a distinction drawn between human time and the eternal time of God (time and eternity) where human time involves a present defined as without extension. This then leads to paradoxes in the measurement and passage of time: 'In what extension do we measure time as it is passing? Is it in the future out of which it comes to pass by? No, for we do not measure what does not yet exist. Is it in the present through which it passes? No, for we cannot measure that which has no extension. Is it in the past into which it is moving? No, for we cannot measure what does not exist' (Augustine, 1998: 236). These paradoxes do not occur in Deleuze's model because past, present and future are all syntheses and because each one is in different ways a dimension of the others, so time is always passing and has

a measure, though not in the sense of clock time or a human sense of that passing, but rather in the stretch of processes. Equally, Deleuze does not have the time as line and time as eternity opposition, relegating both to misleading representations of underlying and more multiple processes. A similar line of argument is perhaps possible in relation to the paradoxes encountered by Aristotle in relation to 'the now' in his *Physics* (see Waterlow, 1982; 1984). Philip Turetzky's reading of Aristotle includes a helpful discussion of the confusion of the now and the present in Aristotle, an identification of the two terms that is avoided in Deleuze's account (Turetzky, 1998: 22–5).

6. These discussions about asymmetry and the arrow of time allow Deleuze's work to intersect with important debates on the arrow or arrows of time in analytic philosophy of time and in discussions in contemporary physics and chemistry. It is important to stress, though, the great contrast in methods between the two philosophical approaches where the analytical studies are very much driven by the physics of quantum theory and thermodynamics, and the way physics 'pick out a direction in time' (Savitt, 1995: 12). It is also important to keep the historical order in perspective here, thus influential philosophical and scientific studies on the arrow of time by thinkers who Deleuze knew well appear quite a while after Deleuze's work in the late sixties (see, for instance, Prigogine and Stengers, 1988: 120–1). Stephen Hawking has a very clear account of the arrow of time and contemporary physics that runs somewhat counter to Prigogine and Stengers, at least in his conviction that it is possible to reconcile arrows of time with laws of physics that are symmetrical by 'not distinguishing forwards and backwards' (Hawking, 1988: 169). It is this latter assertion, in its popular version as given by Hawking, that also stands in opposition to Deleuze's speculative and metaphysical commitment to asymmetry in his syntheses of time. For a very clear account of time's asymmetry and hence how these speculative views can at least be consistent with physics and the second law of thermodynamics, see Lockwood, 2005: 187–92. For a strong philosophical overview of questions around the arrows of time, see Le Poidevin, 2003: 202–18. Jeffrey Bell has a detailed discussion of dynamic systems that makes use of work by Prigogine and Stengers, as well as by Stuart Kauffman, in order to explain the connections between Deleuze's philosophy and the relation between order and chaos in dynamic systems. The arrow of time can be explained as the shift from chaos to order according to the second law of thermodynamics, but what Bell shows is that in dynamic system theory, order can also emerge out of chaos (Bell, 2006: 200–3). Both of these processes are asymmetrical and their coexistence offers an analogy with the multiple forms of asymmetry found in Deleuze's philosophy of time. In *The Universal (in the Realm of the Sensible)*, Dorothea Olkowski makes an original connection between dynamic systems, Bergson's cone of time

and Deleuze's plane of immanence and work on the event. These connections benefit greatly from an effort to rethink them in relation to human sensibility and scales of events (Olkowski, 2007: 219–21).

7. I raise the specific example of Merleau-Ponty because of its closeness to Deleuze's treatment of synthesis and habit on many points, for instance, around the question of the relation between bodily synthesis and habit: '[. . .] habit in general allows us to understand the general synthesis of the body proper' (Merleau-Ponty, 1945: 177). For a good discussion of the closeness and distance of phenomenology and Deleuze's philosophy of time, see Reynolds, 2007: 144–6. For a broader and very helpful discussion of Deleuze in relation to Merleau-Ponty and phenomenology, see Reynolds and Roffe, 2006. For critical remarks on Merleau-Ponty and spacialisation, see Olkowski, 1999: 83–8. These remarks are interesting in this context because Olkowski follows up on her critique of Merleau-Ponty with a study of the roots of Deleuze's second synthesis of time in Bergson's work and in *Matter and Memory* in particular. The originality of this reading lies in its focus on time, memory and the body free of a phenomenological basis (Olkowski, 1999: 109–13).

8. For a radical opposition to Deleuze's speculative attempt to develop a philosophy of time away from objective and subjective grounds, yet consistent with a determining relation to both, see Reichenbach, 1958. In the conclusion to his work on mathematics and space and time, in relation to Einstein and relativity, Hans Reichenbach draws a stark opposition between subjective and objective approaches to time, refuting any conflation of the two, through a study of coincidence: 'It is a serious mistake to identify a coincidence, in the sense of a point-event of the space–time order, with a coincidence in the sense of a sense-experience. The latter is *subjective coincidence*, in which sense perceptions are blended; for instance, the experience of sound can be blended with the impression of light. The former, on the other hand is *objective* coincidence, in which physical things, such as atoms, billiard balls or light rays collide and which can take place even when no observer is present. The space–time order deals only with objective coincidences, and we go outside the realm of its problems in asking how the system of objective coincidences is related to the corresponding subjective system' (Reichenbach, 1958: 286). Reichenbach's anti-Kantian arguments are forerunners of other critical reactions to transcendental philosophy and to its development in Deleuze's work through a purported dependence on subjective grounds and hence mistaken association of objective facts with a subjective condition, for instance, in Quentin Meillassoux's critique of 'correlationism' (Meillassoux, 2006: 50–3). It is interesting, though, to note that Deleuze's critique of objective grounds for repetition is consistent with Meillassoux's arguments for the necessity of contingency.

Endnotes

9. This is a simplification of the question of probability in Hume, notably in its presentation in *A Treatise of Human Nature* and its discussion of unphilosophical probability as derived from general rules rather than experience (Hume, 2009: 99). The main point I want to make here is the connection between Deleuze's account of anticipation and Hume's explanation of the force of belief according to probability: 'The belief, which attends the probability, is a compounded effect, and is formed by the concurrence of the several effects, which proceed from each part of the probability' (Hume, 2009: 94). The compound effect and the concurrence correspond to the habitual contraction as described by Deleuze. For a discussion of Deleuze's interpretation of Hume in relation to the problems of empiricism and belief, see Martin Bell's discussion of Deleuze's transcendental empiricism in relation to Hume (Bell, 2005).
10. It is important to note that Deleuze does not follow Bergson's treatment to the letter in the case of the clock strikes. As in the case of Hume, Deleuze renders the example in a very pure form that elides part of Bergson's scenario (a shift in attention during the clock strikes). Deleuze also sidesteps a key distinction in Bergson's argument drawn between qualitative and quantitative, where the qualitative indicates a synthesis not available to a quantitative treatment (Bergson, 1959: 85). This treatment in terms of intensity rather than number is not present with such force at this stage of Deleuze's argument, partly because, as we shall see in Chapter 5, his account of intensity and internal differences resistant to separation and identification is associated with the third synthesis of time and not the first.
11. An interesting area for further research outside this treatment of time lies in Deleuze's formalisation of Hume's examples. These examples are primarily experiential and often moral; they are connected to investigation of human nature, for instance, in terms of the effects of sarcasm or flattery (Hume, 2009: 102). Deleuze's aim, however, is to give Hume's account of synthesis its widest application possible.
12. Deleuze's discussion of signs of love and lost time is particularly helpful for understanding signs as passage and becoming, rather than indication or token. Signs include their own passing in order to work as signs of things that also inherently pass, such as love and life: 'The signs of love and jealousy carry their own alteration within them for a simple reason: love never ceases preparing for its own disappearance, to mimic its rupture' (Deleuze, 1970: 27).
13. Note that Deleuze swaps his usage of 'structure' and 'system' between books and sometimes even within a particular text. Here, I am using the terms in line with his use of 'system' in relation to 'series' in his account of the third synthesis of time (see Chapter 5, below). This usage is at odds with the stretching of the concept of structure towards a differential system, which occurs in, for example, 'How to recognise

structuralism?' (Deleuze, 2002: 238–69) or in *The Logic of Sense* (LoSe: 48–51).

14. See Daniel W. Smith's work on Deleuze and Leibniz for an illuminating discussion of individuation in Deleuze in relation to Leibniz through the concept of pre-individual singularities (Smith, 2009: 62).
15. This is a micro-level version of the grandfather paradox, on the contradictions generated by time travel if the traveller goes back in time to kill his or her grandfather, thereby eliminating the causal chain making that particular return in time possible in the first place (see, for instance, David Lewis's arguments in 'The paradoxes of time travel': 'Tim cannot kill grandfather. Grandfather lived. So to kill him would be to change the past' (Lewis, 1986: 77).
16. The opposition between a time of succession and a time of eternity where there is no unfolding of processes is central to Augustine's argument, for instance, in relation to the opposition of the Word of God and the spoken word: 'It is not the case that what was being said comes to an end, and something else is then said, so that everything is uttered in a succession with a conclusion, but everything is said in the simultaneity of eternity' (Augustine, 1998: 226). In Chapter 5, we shall see that Deleuze retains the idea of eternity and a sense of simultaneity in his adoption of Nietzsche's eternal return, but that these are reserved for a differential process rather than the immediacy of identity central to Augustine's understanding of God: 'And so by the Word coeternal with yourself, you say all that you say in simultaneity and eternity, and whatever you say will come about does come about' (Augustine, 1998: 226). This coeternity is opposed to Deleuze's idea of coexistence in his treatment of the second synthesis of time, as is the idea of the creation of existence through eternity. For Deleuze, there is coexistence only in a passing away into the pure past and the creation in eternal return is one of difference. Both these processes make the same pass and only difference return.
17. It is questionable whether Deleuze's treatment of case and element, combining Hume and Bergson, is consistent with Bergson's own thought of duration and multiplicity. There is a dialectics of case and element in Deleuze where multiplicity is the result of the interaction, the difference between each term that makes it dependent on the other but not reducible to it. For Bergson, it seems that this mutual dependence of case and element would impose a symbolic order on duration and multiplicity. Thus, Bergson distinguishes real multiplicity, as qualitative and durational, and homogenous and symbolic multiplicity, as quantitative and divisible: 'To conclude, let's distinguish two forms of multiplicity, two very different appreciations of duration, two aspects of conscious life. Beneath homogenous duration, extensive symbol of true duration, an attentive psychology untangles a duration whose heterogeneous moments interpenetrate' (Bergson, 1959:

85). Frédéric Worms argues persuasively that this still demonstrates the necessity of living consciousness for any approach to duration, so in this Bergson and Deleuze would still hold consistent positions (Worms, 2004: 201). This is important because it counters the critique of Deleuze through Bergson in terms of a necessarily subtractive creativity capable of approaching the virtual and duration (Hallward, 2006: 80–1). The dialectics of two multiplicities need not be seen as dependent on an otherworldly scale of cosmological and ontological creation. It is present in the most mundane acts such as counting four strikes of a clock. There is strong evidence for this reading of the importance of everyday distinctions for the expression of duration and intensity in Deleuze in relation to a reading of Bergson; see, for instance, Deleuze's lesson of 21 March 1960 on Bergson's Creative Evolution (Deleuze, 1960: 168–9). For another and more direct reference to the problem of two types of multiplicity, see Deleuze's *Le Bergsonisme* (Deleuze, 1966: 33). John Mullarkey has a counter-reading to this view of multiplicity, based also on a reading of Deleuze in relation to Bergson, where there is an 'ascendancy of the virtual at the expense of the actual' (Mullarkey, 2006: 25).

18. For a discussion of the role and interpretation of signs in relation to Deleuze's philosophy of time, see Keith Faulkner's *The Force of Time*. Faulkner's reading is helpful in understanding the relation between Proust and Bergson in the development of Deleuze's reading of time in relation to signs. In particular, he is good at explaining how signs unlock the philosophy of time through the force of an encounter with intensity: '[. . .] signs are not *time itself*. They merely express its force' (Faulkner, 2008: 20).

19. Levi Bryant has a deep discussion of the importance of the sign as encounter in relation to Deleuze's philosophy of time. Bryant's account of signs is strongly Kantian in its framing but captures the disruptive process at work through signs and the novel form of sensation that it implies (the *sentendium*). Thereby, it moves beyond the letter of Kantian philosophy while absorbing its language, in particular in relation to faculties. In the specific context of this discussion and the important distinction between natural and artificial signs, this Kantian heritage comes out in Bryant's association of artificial signs with language and natural signs with a smooth, that is, non-encounter-like contraction in a habitus. At least in terms of the distinction drawn in relation to the first synthesis of time neither of these associations is correct. This leads Bryant to posit a third type of transcendental sign: 'As a result there must be a third type of sign which is neither natural nor artificial. Although Deleuze does not name it, such a sign would be a transcendental sign' (Bryant, 2008: 100). This need to posit another condition is, in my view, symptomatic of a reading of Deleuze's philosophy of time at odds with the one given here. The natural sign is

already an encounter and a transformation, and the relation to the habitus, that is, to the series contracted in the first synthesis of time, is one of novel metamorphosis. Bryant renders the first synthesis as too natural and thereby downplays the extent to which it is already transcendental. This is because the manner in which the present is a dimension of the past and the future is also downplayed in his interpretation, when compared with the one given here. The multifaceted aspect of Deleuze's philosophy of time and its wide set of mutual conditions are organised hierarchically by a reading that sets out a series of directional transcendental steps rather than a weave of reciprocal conditions. Deleuze's account of the real in relation to time is then directed towards the transcendental: 'Time becomes the dimension of the transcendental, its defining feature, and internal difference is to be equated with temporal difference' (Bryant, 2008: 266). This transcendental is encountered through signs and time is indicated by this special encounter. Against this pure transcendental structure allied to empirical encounters, my reading of Deleuze's philosophy of time insists on a multiplicity of processes all of which articulate transcendental conditions. For Deleuze, time is not the transcendental; it is process determined transcendentally. Time is not restricted to special encounters. In reality, all processes are such encounters, not just some of them.

20. Keith Ansell Pearson gives a very helpful explanation of this relation between the larval subject that is not a conscious subject and a need that is not a lack: 'Deleuze thus insists that underneath the self of action there is to be found those larval selves – molecular selves – that contemplate and that render possible the actions of the subject (Ansell Pearson, 1999: 101). I have eschewed references to Deleuze's later works in explaining the philosophy of time in *Difference and Repetition*; in this case, however, the use of 'molecular' from *Anti-Oedipus* is an illuminating connection since the larval selves work underneath molar entities such as the conscious self in the same way as molecular flows. A question left hanging by this cross reference between the two books remains in the connection between the concept of synthesis and its application to time in the earlier book and the work on syntheses in the later one. My view is that though it is tempting to equate the three syntheses of time to connective, disjunctive and conjunctive syntheses in *Anti-Oedipus*, this move involves many risks of confusion due to the dramatic shift between time and psychoanalysis as the locus for the processes in the two books. The best discussion of these questions is in Lampert, 2006. There is an excellent account of the three syntheses from *Anti-Oedipus* in Eugene W. Holland's *Deleuze and Guattari's Anti-Oedipus: Introduction to Schizoanalysis*. Holland does not refer to the syntheses of time or to *Difference and Repetition* in his analysis, but he does make reference to notions such as past tense and repetition in

Endnotes

such a way as to leave open the possibility of a reading back through the earlier book (Holland, 1999: 34–5).

3 THE SECOND SYNTHESIS OF TIME

1. There is a deep reflection on this problem and many others associated with the three syntheses of time in Jay Lampert's *Deleuze and Guattari's Philosophy of History*. Lampert's interpretation in terms of the paradoxes generating the second synthesis is exemplary: 'Pastness must be a temporal property of the event that does not result after something has happened to the present (just as the future cannot just be presents that have not happened yet). Its pastness must be as much part of the actual event as its presentness is. The event must have its character of being past independently of, at the same time as, its character of being present' (Lampert, 2006: 45). As an exercise in logical close analysis of Deleuze's arguments Lampert's approach has much to recommend it, in particular, given the way in which he analyses some of the most difficult aspects of Deleuze's philosophy of time through illuminating historical references and with the ultimate goal of giving an account of Deleuze and Guattari's philosophy of history. Lampert's general method is to concentrate on the problem of explanation: 'But how does the past's perspective *explain* how, or indeed concede that, the present passes?' (Lampert, 2006: 46). This method is slightly at odds with the approach to Deleuze's philosophy of time given here, because of my emphasis on transcendental conditions rather than logical explanation. Lampert is extremely close to Deleuze's text yet also brings in a form of enquiry that accepts a conception of explanation that privileges a standard logical framework (non-contradiction and so on) where it seems to me that Deleuze's philosophy of time requires a stronger sense of the generative and revealing function of paradoxes, rather than seeing Deleuze as attempting to resolve them.
2. Deleuze's resistance to any explanation through the void or nothingness is at the heart of Badiou's critique of Deleuze's philosophy of time as it is developed in terms of eternal return. Badiou's appeal to a void between events is opposed to Deleuze's eternal return of pure difference, rendered as the return of differential relations that reintroduce the new, transform all actualities, making the present and the same pass, while at the same time affirming pure difference through the same that passes. Badiou's argument therefore has four critical components, each related to the absence of void: Deleuze's thought is of Relation as the One, because there can be no separation between relations (Badiou, 1997: 97); eternal return is the return of difference as only truth governed by a law of the One, because all pure differences must return each time and not a particular truth independent of others (107); chance is reduced to a chance of the One, since all of chance is

drawn together in each event, in one great Event, since there is no void between events (113); Death as transformative event becomes a ubiquitous and necessary principle according with the return of difference, rather than a nihilating passage into nothingness (116). This explains Badiou's plaintive statement on Deleuze's own death as curtailment of their exchange of philosophical letters: 'For me, alas! – contrary to his own heroic conviction, sustained by the incorporation of the One and unity of chance –, death is not, is never an event' (116). Another passage from Badiou's study of Deleuze's philosophy of time captures the full argument better, though: 'And just as truths are singular and incomparable, the risky events where truths find their origin must be multiple and separated by the void' (115). The counter to these criticisms and position statement is that since there is no such void. Any commitment to singular truths is a distortion and false solution to the problems raised by the relations between all events. Events are never separated because they are generated by pure differences expressed differently in all events. Furthermore, the counter-critique to Badiou's dependence on the void is that Badiou cannot at the same time assume that there are rare events and singular truths and a void independent of them, since nothing in principle prevents the void subdividing events, situations and truths further. From this angle the more general criticism is that Badiou's philosophy is either open to an atomisation or to the accusation that his selection of events and truths is contingent and particular. It is one version of events but the presupposition of the void opens up the possibility of others.

3. Discussions of the concept of void in relation to time are somewhat rare when compared with other topics, such as the arrows of time. However, there is an interesting discussion of the void in relation to Aristotle, Parmenides and Leibniz and time in Robin Le Poidevin's *Travels in Four Dimensions* (Le Poidevin, 2003: 18–29). Le Poidevin's arguments are particularly useful in the context of doubts about Deleuze's method in relation to metaphysical assumptions because Le Poidevin rejects Aristotle's arguments against a spatial void on empirical bases: 'However, the fact remains that Aristotle's arguments against the void are weak in that they rest on mistaken physics' (34). If Deleuze's arguments have some resistance to this kind of dismissal it is because they do not rest on physics, whether weak or not; they are instead speculative proposals within a consistent metaphysics. Of course, this does not make those arguments immune to a different kind of appeal to physics, that is, that the speculative philosophy is contradicted by physical theories. Le Poidevin's analysis of the question of temporal voids has a further critical interest for the arguments set out here, since it couches the question of the void in relation to the wider problem of time without change, something that my reading of Deleuze seems to disallow by definition through the claim that time is

(made with) process. If there can be time without change, time without any process, then Deleuze's philosophy of time is flawed in some fundamental way. There is a further interest in Le Poidevin's study in this direction, since he sees Leibniz's principle of sufficient reason as one of the stronger arguments for discounting time without change. This chimes with Deleuze's own dependence on such arguments (as charted in Smith, 2009).

4. I have discussed the challenge presented by a mounting number of dead souls elsewhere through a study of Deleuze's debt to Charles Péguy's *Clio*; see Williams, 2008a, 117–19; 2009: 142–9.

5. For an extensive study of the history of the philosophy of time in conjunction with a political and historical survey of its ramifications, see Alliez, 1996. This study is particularly interesting in relation to Deleuze's distinction drawn between foundation and founding in his own philosophy of time, because Alliez takes the distinction as a starting point for an understanding of the critical relation between concepts of time and the worlds and objects constructed along with them (Alliez's book has a generous and illuminating foreword by Deleuze). Like Deleuze, Alliez is careful not to separate foundation and founding as if they belonged to two independent times. It is, rather, that we can only grasp the constructive and destructive power of time when we understand that it has to be both: 'Put this way, the question of time led me to an enquiry that could be "foundational" only by confronting an unfounding (*effondement*) – to use Deleuze's word – dimension, which to my mind would be inseparable from the critical labor of thought from its and upon its own history' (Alliez, 1996: xxi). Alliez's book is of further interest here because it bridges between the earlier ideas about time in Deleuze, such as founding and ungrounding, and later more spatial concepts, such as reterritorialisation and deterritorialisation, from the work with Guattari. His reading therefore provides a critical contrast to mine, due to my more strict focus on *Difference and Repetition* and on *Logic of Sense*. As such, Alliez's work can be studied and aligned more closely with Jay Lampert's approach (Lampert, 2006). What Alliez shows is that these apparently spatial concepts can also be thought of in relation to time because, for instance in Aristotle's work on the 'now', time has to be 'placed' (Alliez, 1996: 22–3). Note that there are strong contrasts between this reading and Lyotard's work on time and Aristotle in *The Differend*. This is because Lyotard's conception of time is developed from a different conception of the event than Deleuze's and a different place for language in relation to time. For Lyotard, the event as break, as an essential yet inexpressible manifestation in language and sensation, provides the frame for thinking about time (Lyotard, 1988: 74–5), whereas for Deleuze, time is made by an essential plurality of events where the concept of break, though present, never comes to be the founding concept. This contrast can

be partly explained by Lyotard's constant concern with the dangers of metaphysics of time when compared with Deleuze's much more sanguine speculative and constructive approach to metaphysics (Lyotard, 1988: 74; see also Wood, 2007: 188–91 for a reading of Lyotard's aesthetics read through time and the event). The contrast can also partly be explained by Deleuze's stronger sense of the multiplicity of times and their separation from a narrower subjective and theological frame, when compared with Lyotard's deep yet critical attachment to an Augustinian ground for thinking about time, flesh and transcendence. This is a theme Lyotard returned to in his last works on Augustine, touching on many of the ideas and problems found also in Deleuze, such as memory and forgetting, but from a more condensed yet perhaps more powerful sensibility and experiential context (Lyotard, 1998: 55).

6. This discussion of soil and sky is mirrored by a similar but deeper discussion in *Logic of Sense* through the kindred concepts of height, depth and surface; see LoSe: 127–33. Of course the two treatments should not be conflated, but they reflect similar concerns and problems.

7. For a nuanced account of Cartesian foundationalism that allows for a hypothetical method in Descartes, see Schmitt, 1986: 505–8. However, even Schmitt's account of hypothetical deductions avoiding standard foundationalism based on indubitable intuitions leads to an opposition between Cartesian methods because on this account we still need a 'two tier' foundationalism of hypothetical deductions and indubitable intuitions to reconstruct the full Cartesian model.

8. Deleuze's need to go beyond Hume's account of the association of ideas, despite Deleuze's appeals to Hume's account of the synthetic power of the imagination in *Difference and Repetition*, can be traced to the 'qualities' giving rise to association according to Hume: 'The qualities, from which this association arises, and by which the mind is after this manner conveyed from one idea to another, are three, viz. RESEMBLANCE, CONTIGUITY in time or place, and CAUSE and EFFECT' (Hume, 2009: 13). Were these qualities taken to be accurate by Deleuze, his whole philosophy of time would collapse through a contradiction between this explanation of association and the syntheses of time which explicitly work against each one of the qualities. Association of this kind must therefore be restricted to active memory by Deleuze and, even then, they would only offer an incomplete account of the operation of memory. This is another reason why Deleuze moves beyond his early work on Hume and into Bergson's work on memory and duration. Hume provides methodological insights and a general empirical frame, but the detail has to come from Bergson. A similar argument is made by Jeffrey Bell in his reflections on the shift from Hume to Bergson (Bell, 2009: 54–6). For Bell, the importance of

Hume is in a 'problematisation' of the actual which is then 'extended', for Deleuze, in Bergson's work (56) However, in order for this problematisation to be taken seriously we have to discard many aspects of Hume's work taken to be central. Chapter VI of Deleuze's very early book on Hume, published fifteen years before *Difference and Repetition*, already shows this kind of distinction in Deleuze's thought around Hume's associationism, in particular, where Deleuze argues that the principles of association are subordinate to those of passion (Deleuze, 1953: 138).

9. For a strong study of these paradoxes see Lampert, 2006: 31–53. For an excellent interpretation of Deleuze through the role of paradoxes in his work, see Montebello, 2008a. Pierre Montebello's reading of Deleuze is one of the most original yet faithful through its work on a broader set of paradoxes at work in the main concepts of his philosophy. The parts of this work most directly relevant to questions in the philosophy of time can be found in the closing sections of Montebello's book where he demonstrates the manner in which Deleuze reads Bergson alongside Spinoza. This is a very original and fruitful take through questions of appearance (228–32). In terms of Deleuze's arguments for the pure past through the distinction between the general and the particular, Montebello provides a subtle answer to a possible criticism of Deleuze that a particular perceived image is stronger than any virtual one in the pure past. On the contrary, the perceived image is 'elimination' through representation of a wider and richer general set; it is an impoverishment rather than a whole (243). Again, the roots of Deleuze's arguments here can be traced back to his earlier work on Bergson in *Bergsonism*: 'It is not such and such a region that would contain such and such elements from the past, such memories, in opposition to another which would contain other ones. It is distinct levels, each one of which contains all our past but in greater or lesser states of contraction. It is in this sense that there are regions of being itself, ontological regions of the past "in general", all coexisting, all "repeating" one another' (Deleuze, 1966: 57).

10. This criticism of Deleuze through a connection to Bergsonian mysticism is made by Peter Hallward: 'Bergson concludes that our only option is "to follow the lead of the mystic". Only mystics, or people who become like mystics, can live in a manner adequate to the living that lives through them. By the same token, only such people, or similar people, can endure what Deleuze will call a virtual *event*' (Hallward, 2006: 23). This is a fundamental narrowing of Deleuze's position, for instance on destiny, but also on the event, where events and destiny are not restricted to people, let alone to a group of people. Furthermore, any adequate living with events and with time takes place through actual selections that owe nothing to mystical experience but rather to a very practical experimentation. Hallward consistently misses this role

of the actual, for instance in the dimensions of time, where the actual living present is either the prior time or a dimension, but never completely missing. Here is the strongest statement of Hallward's critique of Deleuze's philosophy of time and of its supposed baleful results on Deleuze's politics: 'First of all, since it acknowledges only a unilateral relation between virtual and actual, there is no place in Deleuze's philosophy for any notion of change, time or history that is mediated by actuality. In the end, Deleuze offers few resources for thinking the consequences of what happens within the actual existing world as such. Unlike Darwin or Marx, for instance, the adamantly virtual orientation of Deleuze's "constructivism" does not allow him to account for cumulative transformation or novelty in terms of actual materials and tendencies' (162). When Deleuze's philosophy of time is studied closely it becomes clear that relations are bilateral, but asymmetrical. It also becomes clear that all notions of process, time and history are mediated by the actual, that is, in relation to the actual living present. The first synthesis of time and its relations to the other syntheses therefore provide strong resources not only for thinking through what happens in the actual world, but also for thinking that through in relation to the past and to the future. Indeed, a concern to do this is clearly at work in Deleuze's remarks on Marx and history, and on biological evolution, in *Difference and Repetition*. His conception of destiny is therefore not mystical at all and certainly not aimed at a virtual realm. Destiny is actual and requires actual selections in relation to it, but it is not only actual and those selections will have virtual effects. Transformation and novelty in the actual present are a key and repeated element of Deleuze's philosophy of time. Furthermore, when Hallward lists a series of contemporary political concerns that Deleuze's philosophy is unable to register or to respond to, such as defence, coordination and commitment, these are important aspects of Deleuze's awareness of our situation within actual and virtual processes demanding such manoeuvres. Hallward could be right about the nature of the political challenges we face, but this does not imply that Deleuze's philosophy of time is not a powerful resource for responding to those challenges. That conclusion can only be met by giving a lopsided account of his philosophy.
11. Giorgio Agamben has given us the most original and detailed reading of 'Immanence: a life . . .' This reading is not, though, consistent with the philosophy of time as outlined here, because it reintroduces a transcendent and eternal realm in relation to Deleuze's ontology of immanence by defining the virtual as absolute, a move that sits uneasily alongside the intricate interdependence of syntheses of times given in *Difference and Repetition* (Agamben, 2000: 224). For a fuller version of this argument, see Williams, 2010.
12. For an excellent discussion of multiplicity in Deleuze in relation to

Bergson, but also taking great care to draw out the differences between them, see Miguel de Beistegui's *Truth and Genesis*. His reading of multiplicity in relation to problems is particularly effective in explaining how a multiplicity is virtual (de Beistegui, 2004: 248–9). Following DeLanda, de Beistegui argues that Deleuze's concept of multiplicity can be best understood in relation to physics and to mathematics (DeLanda, 2002: 71–3). I have reservations about this argument in its application to Deleuze's philosophy of time, in particular where the claim is made that mathematics gives the definitions within which Deleuze's philosophy operates, and it is important to realise that de Beistegui goes further than DeLanda by showing the non-mathematical part of Deleuze's argument (de Beistegui, 2004: 266).

13. The problem of a whole life is related to the problems generated by problems of parts and wholes and their implications for the broader problem of time without change as discussed by David Lewis in his essay on time travel (Lewis, 1986: 68–9). It is interesting to contrast Lewis's solution where enduring things are wholes composed of temporal parts, thereby allowing for change between the parts, with Deleuze's concept of series where change is the essence of any series or whole. Time without change is not therefore a deep problem for Deleuze, contrary to Lewis and Aristotle, since in some sense there is only change and the incomplete illusion of unchanging parts. However, what is a problem for Deleuze is the relation between parts and wholes, in the sense that his account seems to disallow the very existence of distinct parts and to imply an undetermined whole. It is interesting to note that discussion of time without change very quickly sets aside the possibility that time is only continuous change in order to derive the problem. This is partly because change is thought of in terms of cause, where change is the effect of an earlier cause. There is therefore a great distance between Deleuze's non-causal account of a time of ubiquitous and continuous change and, for instance, Shoemaker's assumptions in his treatment of time without change: 'If time is dense or continuous, of course, we cannot speak of a change as being caused by the state of the world at the immediately preceding instant, for in that case there is no immediately preceding instant. But I think it is rather commonly supposed that if an event E occurs at time t and is caused, then, for any interval I, no matter how short, that begins at some time prior to t and includes all the instants between that time and t, the sequence of the world states that exist during I contain a sufficient cause of E' (Shoemaker, 1993: 75). Not only does Deleuze's philosophy deny the assumptions about cause, limited events, the existence of time t, the concept of discretely prior in time and the notion of the sequences of the world, his speculative metaphysics is constructed to provide a thorough critique of the appeal to common supposition with all its presuppositions about time.

14. In his 'Essai sur les données immédiates de la conscience' Bergson's critique of Kant's philosophy of time is unequivocal and severe. He accuses Kant of confusing time with space by taking time as a homogeneous medium which therefore forces him to miss the way in which durations are internal to one another, that is, durations coexist, whereas according to Kant's model they must be positioned separately and as external to one another like spatial objects (Bergson, 1959: 149).
15. The legacy of this fusion of the human and the animal, against Mill, can be found in the concept of becoming-animal developed by Deleuze with Félix Guattari (see Ansell Pearson, 1999: 179–80). For a different reference and one that makes a direct reference to pigs, see Deleuze's *Logique de la sensation* (Deleuze, 1981: 19–22). Deleuze returns to the 'problem of foolishness [*bêtise*]' in chapter III of *Difference and Repetition*, also referring it to animals and their lack of *bêtise* (DRf, 196–8).
16. This reference to Sartre and Genet is tentative, since there could be other associations of criminal and saint that Deleuze is alluding to. For further connections between Deleuze and Genet see a short footnote in *Cinéma-1* (Deleuze, 1983: 191) as discussed in Conley 1997 and Frichot 2007.

4 THE THIRD SYNTHESIS OF TIME

1. Keith Ansell Pearson is one of the few commentators to have followed this shift from Descartes to Kant in the philosophy of time in detail (Ansell Pearson, 1999: 101–3). Other commentators tend to separate out Kant and Descartes, reserving the discussion of time for the former and the subject and thought for the latter; see, for instance, Bryant's strong discussion of thought in Descartes and Deleuze (Bryant, 2008: 135–41).
2. There has been very little work on Deleuze and Sartre in relation to time. Yet it is striking how Sartre in the 'Temporality' chapter of his *Being and Nothingness* covers very similar ground to Deleuze. For an example of this critical overlap and of its potential for further research see, for instance, Sartre's own discussion of Descartes, Kant and Bergson: 'Leibniz, reacting against Descartes, Bergson, reacting against Kant, have only in turn wanted to see a pure relation of immanence and cohesion in temporality' (Sartre, 1943: 173). These pages by Sartre suggest a potential Sartrean critique of Deleuze's philosophy of time and its own albeit more complex commitment to continuity, immanence and relations in the syntheses of time.
3. Deleuze takes two points from Hölderlin's reading of Sophocles from his essays on theory. First, the caesura is crucial for understanding the role of time: '[. . .] there becomes necessary *what in poetic meter is called cesura* [. . .]' (Hölderlin, 1988: 102). Second, in this caesura the subject

exists only as a moment in a time the subject becomes passive to, that is, separated from any ends or teleology set before the caesura: 'Inside it man forgets himself because he exists entirely for the moment, [the god] forgets himself because he is nothing but time; and either one is unfaithful, time, because it is reversed categorically at such a moment, no longer fitting beginning and end; man, because at this moment of categorical reversal he has to follow and thus can no longer resemble the beginning in what follows' (108).

4. The most extensive discussion of founding in relation to Plato is by Dan Smith. Smith's argument is important because it opposes foundation and 'ungrounding'. In Deleuze's philosophy of time it is important to add the process of founding through the pure past to the operation of foundation. Founding would be a process that corresponds to an ungrounding in Smith's reading because it does not set up a secure foundation, but rather sets levels within the pure past such that a passing present neither passes into a pure chaos nor becomes a settled and final foundation for future events. This is why I have some qualms about Smith's use of a chaos–'universal collapse' dichotomy: 'Far from being a new foundation, the simulacrum allows no installation of a foundation-ground; rather, it swallows up all foundations, it ensures a universal collapse, an "un-founding" (*effondement*), but as a positive event, a "gay science"' (Smith, 2006: 113). In terms of Deleuze's philosophy of time, the processes of time, working through simulacra, install a third determination between foundation and chaos: foundation-founding-chaos. In fact, I would argue that the two extremes are illusions: there is no pure chaos, nor foundation, only levels of founding that can be obscured by falsifying images of chaos and of the absolute.

5. Jacques Derrida also discusses Hamlet and the statement that time is out of joint in his reflections on time (and much more) in *Spectres of Marx*. Derrida's study has a very beautiful series of reflections on translating Shakespeare's work that draw out the difference in approach between Deleuze and Derrida around the pace of language in relation to the philosophy of time (Derrida, 1994: 19–23). Tamsin Lorraine has studied Deleuze and Derrida's different takes on time and on Hamlet in her chapter 'Living a time out of joint'. She concludes with a distinction between the methods of the two thinkers on time, where Derrida stages encounters with an 'impossible time' and Deleuze creates concepts that situate sense in 'the fractured perspectives of a durational whole'. This fracturing and perspectives are what we find in the place of the subject in Deleuze's third synthesis of time. This fracturing is seen as consistent with Derrida's encounters with time since he and Deleuze 'shift to a non-representational view of time – one that could reveal the intensive processes of becoming rather than reifying the equilibrium states of its continual unfolding – could reveal a different kind of subjectivity, one able to do justice to the other, to the tensions

between words and things, and to life, despite the fractured time of its evolutions in learning to live' (Lorraine, 2003: 45). In another discussion of Deleuze and Derrida, Leonard Lawlor draws out a different connection between them, in this case their displacement of the present away from presence, place and the instant. The great merit of Lawlor's reading is in his reading of *Logic of Sense* alongside Derrida's work on the trace which allows a thinking of time as 'becoming' and 'beyond being' (Lawlor, 2003: 77).

6. It is interesting to contrast Deleuze's use of asymmetry in his third synthesis of time and David Lewis's discussion of the asymmetry of past and future in relation to counterfactuals in 'Counterfactual dependence and time's arrow': 'In short, I suggest that the mysterious asymmetry between open future and fixed past is nothing else than the asymmetry of counterfactual dependence. The forking paths into the future – the actual open and all the rest – are the many alternative futures that would come about under various counterfactual suppositions about the present. The one actual, fixed past is the one past that would remain actual under this range of suppositions' (Lewis, 1986: 38). We can speak of the future depending on the present but not of the past depending on the present. So a counterfactual such as 'if this were the case, then that would be the case' can be true with respect to a range of possible worlds for the future, but only with respect to the actual world for the past. The interesting part is that for Deleuze, we have forking paths for the past and for the future, so we can speak of the past depending on the present and we can have a counterfactual about the past that is true for possible worlds other than the actual one. Yet for Deleuze, time is still asymmetrical because possibility and actuality are not sufficient for thinking about time. There are alternative pasts for Deleuze, as opposed to Lewis's view that it is fixed, but the nature of that range is not in possible past worlds varying from the actual one in this or that identifiable feature, but rather in many virtual worlds with different levels in the pure past that are selected in the present, bringing about different series and orders of time, and depending on the different intensities that can be associated with the same actual events. A good way of understanding this is through counterfactuals. From Deleuze's philosophy of time, the counterfactual 'if it were the case that A, then it would be the case that B' is never fully decidable in terms of its truth, and indeed not particularly interesting with respect to thinking about the world because all the conditions the counterfactual depends upon change with the syntheses of time, constantly and according to many different processes. It is not that we have many possible worlds and one actual one, but rather that worlds are intrinsically manifold; not many worlds, but the many in worlds and a many that is not determinable through the concept of the possible.

7. See Hölderlin, 1988: 103–5.

Endnotes

8. It is at this point perhaps that we could begin a long and fruitful discussion of the overlaps and contrasts between Deleuze and Ricoeur on time, for example around the role of the event as that which, in Ricoeur's words that come so close to Deleuze's, 'gives [a plot] the dramatic form of a change in fortune' (Ricoeur, 1984: 225). See Williams 1996 for a longer discussion of Deleuze and Ricoeur's accounts of time and narrative.
9. Levi R. Bryant has a very good discussion of the problem of dogmatism in his *Difference and Givenness*. However, Bryant's argument depends on associating dogmatism and speculative philosophy, and contrasting the two with Deleuze's transcendental empiricism (Bryant, 2008: 176). I disagree with this move, because it confuses two types of assertion: a dogmatic assertion such that a given premise is true independent of experience or scientific discovery, and a speculative assertion that is not based on science or phenomenology, or common sense, yet remains provisional and speculative, that is, open to change if the evidence changes. More generally, Bryant offers a contrasting reading of Deleuze's philosophy of time to the one given here. This is because it treats the relation to time as essentially transcendental and critical whereas the interpretation given here accepts that the form of time is deduced transcendentally, but that time itself is process, that is, syntheses of time. So we get a separation in Bryant's work between transcendental time as condition and genesis, the essence of which is to be found in intensity, and we find Bryant using expressions such as 'unfolding in time' where sensation and time as treated as separate: 'If I am able to conceive an increase or decrease of a sensation a priori, then this is because I can conceive it as unfolding in time as becoming' (Bryant, 2008: 243). My view is that time as synthesis is genesis as process, and processes are time, as opposed to Bryant's phrasing where different syntheses of time account not for processes but things (for instance, persons) (252). In short, Bryant's approach to Deleuze is still too Kantian and does not read Deleuze's philosophy of time enough on its own speculative grounds where time is not a 'transcendental field' (266) but rather a multiplicity of processes the forms of which are deduced transcendentally and whose relations are in part transcendental, in the sense of condition used here.

5 TIME AND ETERNAL RETURN

1. For an extended discussion of questions of priority in Deleuze's philosophy of time see my debate with Jack Reynolds in *Deleuze Studies* (Reynolds, 2007; 2008; Williams, 2008b).
2. There is a rewarding and detailed reading of Deleuze's interpretation of Nietzsche's doctrine of eternal return in Elizabeth Grosz's book on time, politics and evolution, *The Nick of Time: Politics, Evolution and the*

Untimely. Grosz explains how repetition, difference and affirmation are interconnected in Deleuze's account (Grosz, 2004: 141) but more importantly she also demonstrates how central eternal return must be to Deleuze's philosophy of time as a philosophy of becoming. The most significant aspect of this demonstration for the discussion here is in the insistence that affirmation is a necessary moment in the eternal return of difference, thus explaining why actual living presents play a central role in the return of difference: 'The affirmation of returning forever is the affirmation of the very being of becoming, of what it is that becomes: returning is thus the highest affirmation of the power of time itself, the "life" of time' (143). In addition, Grosz also explain why interpretations of Nietzsche's eternal return as eternal return of the same identities do not stand up to scrutiny once they have to account for the relation between eternal return and becoming (143).

3. See Ansell Pearson, 2002: 197–205. Ansell Pearson gives a rich account of eternal return in relation to a reading of Nietzsche and Kant. This is in contrast with the more strongly transcendental and Kantian interpretations found in Bryant, for instance, whose argument plays down the importance of Nietzsche (Bryant, 2008: 221–2). For Ansell Pearson's interpretation of Deleuze's three syntheses in relation to evolution and Bergson, see Ansell Pearson, 1999: 99–103.

4. Deleuze's reading of Nietzsche's eternal return is deeply indebted to Pierre Klossowski's work on eternal return in his *Nietzsche and the Vicious Circle*. Klossowski's interpretation is in fact much more detailed and deeper than Deleuze's, partly because Deleuze is using Nietzsche's ideas within his own construction of time and partly because Klossowki is wrestling with eternal return as the problematic key to a comprehensive and revolutionary account of Nietzsche's life and oeuvre. It is beyond the scope of the analysis of time given here to consider the full range of overlaps and critical contrasts between the two readings, a very promising research project in itself. However, an extraordinary line of thought relevant to Deleuze's philosophy of time can be found in Klossowski's remarks on will and action in relation to the irreversibility of time: 'For these fragments also suggest a transfiguration of existence which – because it has always been the Circle – wills its own reversibility, to the point where it relieves the individual from the weight of its own acts *once and for all*' (Klossowski, 1997: 69). I have quoted this passage because it shows the continuity of Klossowski and Deleuze's work through the 'once and for all' and 'transfiguration' that they both insist upon in relation to the transformations demanded by eternal return, but also because Klossowski introduces an element much harder to reconcile with Deleuze's philosophy of time: reversibility (as opposed to Deleuze's insistence on asymmetry). So for Klossowski it is the creative will itself that frees its acts from the past by willing reversibility – a willing incompatible with Deleuze's account

of time – whereas for Deleuze it is the passing of the same and eternal return of difference that accomplish this freedom without a necessary appeal to will or to affirmation (perhaps in contrast to Deleuze's work on Nietzsche in *Nietzsche and Philosophy*). It is interesting to note the much greater depth and subtlety of both these interpretations when compared with Heidegger's reading of eternal return as 'permanentizing of the unstable' (Heidegger, 1991: 212) or Badiou's 'eternal return of the One' (Badiou, 1997: 113), both of which encompass eternal return within an ontological category (Being or the One), thereby missing the multiplicity of processes in relation to time, willing, creating, passing, perishing, returning and not returning folded into Nietzsche's perplexing aphorisms.

5. There is a strong contrast between Deleuze's study of time as eternal return and the more overtly existential reflections on time, eternity and existence in Kierkegaard. It can seem that Kierkegaard offers a richer examination of the relation between time and eternity by, for instance, adding love to anguish in the list of affects caught on the cusp between time and eternity: 'But the absolute τέλος is present only when the individual relates himself absolutely to it, and they cannot, as an eternal happiness relating itself to an existing person, possibly have each other or tranquilly belong to each other, that is, in temporality, in the same way as a girl and a young man can very well have each other in time because they both are existing persons. But what does it mean that they cannot have each other in time?' (Kierkegaard, 1992: 397). However, this study of the relation between love and eternity, when viewed from Deleuze's statement of the immanence of eternal, leads to a 'leap' into transcendence, to a love that would itself be transcendent and outside the temporal world. It is this leap that Deleuze criticises in Kierkegaard's approach to events, in *Difference and Repetition* (DRf, 17). These remarks and suggestions are far from conclusive, though, and the study of Deleuze's relation to Kierkegaard on time and repetition, in relation to eternal return, remains a deeply promising area of research (see, for example, Battersby, 1998: 176–98).

6. See Nathan Widder's *Reflections on Time and Politics* for an argument against interpretations that reduce the appeal to will in eternal return to a form of solipsism in relation to death (Widder, 2008: 98–9). Widder is particularly good at demonstrating the interconnections of Deleuze's thought, thereby resisting any readings that settle too quickly on a received view of only part of Deleuze's work.

7. The main work in the literature analysing life, death and time in Deleuze is Ansell Pearson's *Germinal Life* (Ansell Pearson, 1999). Ansell Pearson brings eternal return into his analysis through a study of Deleuze's work on destiny in Zola. Eternal return transforms the death drive into an affirmation of life: 'Death effects the transmutation

of instincts, turning death against itself, so creating "instincts" that do not involve simply a repetition of the same but are allowed to live and grow germinally' (Ansell Pearson, 1999: 118). Pierre Montebello also makes a series of interesting remarks that draw the work on death and the third synthesis of time into Deleuze and Guattari's work on life in *Anti-Oedipus* (Montebello, 2008: 189–91).

8. Catherine Cazenave traces the roots of this thought of chance back to Deleuze's *Nietzsche and Philosophy*. This is a particularly effective reading since it also shows the birth of Deleuze's interpretation of eternal return in relation to will to power and affirmative and destructive forces, ideas that take a far less important role in *Difference and Repetition* than they do in the Nietzsche book. In turn this allows for a study of the importance of values and revaluation in relation to chance and eternal return: 'Will to power and the dice throw are thus values of affirmation because they are inscribed in the movement of eternal return, where only an active dynamism rules' (Cazenave, 2006: 108). Cazenave's reading completes this analysis of chance through the important Deleuzian study of Mallarmé's *Un coup de dés*, a connection that is also crucial to understanding the differences between Badiou and Deleuze on chance and time. Cazenave's argument is critical in relation to Badiou's because she notes more closely how Deleuze's focus on Nietzsche allows him to distance himself from what he sees as a negative aspect of chance as treated by Mallarmé: 'Deleuze's first argument concerns the preponderance of negative values in Mallarmé's poem' (111). Badiou discusses the 'Nietzsche or Mallarmé' question in *Deleuze: the Clamour of Being*. Badiou comes down on the side of Mallarmé because he sees Nietzsche and therefore Deleuze's treatment of chance as still a thought of the One rather than a thought of multiplicity: 'For Deleuze chance is the play of All, always replayed as it is' (Badiou, 1997: 115). From the point of view of the interpretation of eternal return and multiplicity given here, this is a rather perverse reading of both concepts, neither of which can be interpreted as leading to a thought of the One with a reduction of their inner and irreducible multiplicity. For a possible response to this critical answer to Badiou, see his discussion of Deleuze on multiplicity, where he argues that Deleuze misunderstands set theory and the concept of multiplicity that comes out of it: 'Deleuze routinely argues that multiplicities, unlike sets, have no "parts". This is indeed what, in my view, explains the fact that the opposition between sets and multiplicities takes place under the aegis of the One' (Badiou, 2004: 76). In relation to time and multiplicity, it is correct to say that time has no finally discrete parts, but Badiou overstretches this remark into the idea that it is the One. There are parts, but these cannot legitimately be separated from a multiple process irreducible to any One. The critical route via sets is really a side issue that stops Badiou from understanding the

properly Deleuzian idea of multiplicity and of a multiplicity of interacting and irreducible processes.

9. In an essential article on simulacra, Deleuze and Plato, Dan Smith traces the concept of the simulacrum in Deleuze's work in relation to his overturning of Platonism. The article is important in relation to eternal return because it explains the difference between Nietzsche and Plato on return and circles. The most important point made by Smith for an understanding of the role of simulacra in series as carriers of pure difference is that simulacra do not function in a mimetic relation to anything independent of them: 'In an inverted Platonism, all things are simulacra, and as simulacra they are defined by an internal disparity' (Smith, 2006: 103). Disparity is pure difference here. This is what allows it to be consistent with eternal return, that is, not to be defined in terms of the same except as the same that always returns: difference.

10. For a different approach to the new than the one adopted here, see Smith, 2007. Dan Smith's tracing of the conditions of the new in Deleuze in relation to calculus and Leibniz is not only outstanding but also challenging for this work, since it gives no prominence to Deleuze's work on time for the determination of the new in Deleuze. This presents two difficulties through the critical question of whether Deleuze's work on calculus should be taken as a starting point for his work on time, as opposed to my focus on synthetic processes, and through the related question of whether the role of singularities in the philosophy of time should be understood through a mathematical understanding of the term: 'The singularities of complex curves are far more complex. They constitute those points in the neighbourhood of which the differential relation changes sign, and the curve bifurcates, and either increases or decreases' (Smith, 2007: 12). My reservation about the mathematical model is its dependence on an opposition between ordinary and singular points. In terms of Deleuze's philosophy of time, there are no ordinary points in ordinary time, since the processes of time are all dependent on multiple singularities and their relations (in the living present, in the pure past, in eternal return and in the caesura that come with the new). In that sense, then, at least for the philosophy of time, my view is that the new is better defined in a more formal metaphysical manner. So I would rephrase the following sentence from Smith's work, avoiding the terms 'ordinary', 'constant' and 'perpetual': 'Every determinate thing is a combination of the singular and the ordinary, a multiplicity that is constantly changing, in perpetual flux' (12). The version closer to Deleuze's account of time would be: Every determinate thing is a combination of singularities, forming a multiplicity that is changing in multiple ways according to the syntheses of time and led by the work of dark precursors and the eternal return of difference, the eternal return of the new.

11. For a detailed interpretation of Deleuze's work on dynamic systems, see Bell, 2006: 200–10. This interpretation is somewhat in contrast to the one given here through the scientific approach to systems in Bell. This does not necessarily mean that the two readings are in opposition, but rather that in terms of the role of simulacra and series in understanding Deleuze's philosophy of time and eternal return, in particular, a more formal approach is more appropriate, whereas in charting Deleuze's relations to contemporary science, a closer comparison of his concepts with those of science is more fruitful.
12. Jay Lampert gives a strong reading of the role of dark precursors in the third synthesis of time. He is critical, though, of Deleuze's philosophy of time in *Difference and Repetition* because 'it does not adequately solve the problem it was introduced to solve. Showing how temporal coexistence can be penetrated requires more emphasis on actual historical events than DR calls into play' (Lampert, 2006: 54). As counter to this critical point I read *Difference and Repetition* as replacing problems of coexistence with manifold interpenetrating processes, so the problem is not well posed if set according to a coexistence and actual date dichotomy. Each actual date can be analysed in terms of the manifold processes it designates and that stand as a condition for it. Lampert comes close to a similar analysis in his account of the different times as 'differentiators' of one another (65). However, this notion is not fully developed and is treated under the idea of each time providing 'a meta-theory of time' whereas Deleuze's idea of dimensions means that all the times together give a complete account of the ways in which past, present and future are conditions for one another.
13. Dan Smith explains the contrast between questions and problems in this context and gives an important situation of the problem in terms of spatio-temporal coordinates rather than a more simple causal chain. The problem is not 'What caused?', but rather 'Who? How? Where? And When?' (Smith, 2006: 111). In the shift from a Platonic transcendent philosophy to an immanent one determined by Deleuze's take on Nietzsche's eternal return, philosophy is returned to a practical and experimental wrestling with problems rather than a solving of them through definitive answers: 'Deleuze's pluralist art does not necessarily deny essence, but it makes it depend in all cases upon the spatio-temporal and material coordinates of the problematic Idea that is purely *immanent* to experience, and that can only be determined by questions such as Who? How? Where? and When? How many? From what viewpoint? And so on' (111). The study of Deleuze's philosophy shows how his work on time complements this practical problematic by revealing time as a network of processes necessarily and productively generating problems. For a detailed argument on mathematics, physics and time in Deleuze, developed in relation to Bergson's critiques of geometrical mathematical models in relation to duration, see Olkowski, 2008.

Olkowski argues that Deleuze is closer to theories of relativity than Bergson but without adopting them wholesale and instead retaining important metaphysical elements such as the concept of image: 'The problem, for Deleuze, is to undo Bergson's antidote to the power of modern science, including the theory of relativity, and to rid ourselves of ourselves, to demolish ourselves, not only our perception and action, but especially our affective states. The image is the mechanism through which this is to be accomplished' (Olkowski, 2008: 15).

6 *TIME IN* LOGIC OF SENSE

1. Brian Massumi has charted the importance of affect and its relation to time over many works. His *Parables for the Virtual: Movement, Affect, Sensation* gives a number of illuminating case studies of the relations between time and affect. For instance, he demonstrates the connection between affects, language and events in a counter to strictly linear accounts of time: 'In this case, *suspense* could be distinguished from and interlinked with *expectation* as superlinear and linear dimensions of the same image-event, which is at the same time an expression event' (Massumi, 2002: 26). A further benefit of Massumi's treatment is his clarity on the definition of affect in Deleuze, where affect is distinguished from emotion and instead defined strictly in relation to intensity, thereby avoiding an overly psychological reading of affect (27).

2. This location of multiplicity in individuation and event provides a counterpoint to Alain Badiou's imposition of metaphysics of the One on Deleuze's philosophy. Badiou consistently reads Deleuze's philosophy of the event in *The Logic of Sense* as a single *and same* great Event for each event (Badiou, 1997: 20). However, Deleuze's philosophy of time and its dependence on individuation and irreducible multiplicity means that it could never be the same Event. On the contrary, Deleuze's point is that everything connects but only on condition that the connection or participation is not through the same but only through difference, so not through the One but through an irreducible many. Badiou's argument therefore confuses the concept of whole in Deleuze with the metaphysical concept of the One. Deleuze's philosophy of time is an attempt to show how the real is whole but never One. Miguel de Beistegui gives a detailed and deep explanation of the role of multiplicity and individuation in his Truth and Genesis. He avoids the reduction of multiplicity found in Badiou's reading and instead charts the steps from virtual Idea to individual through a series of immanent geneses: 'The relation is indeed genetic, even ontogenetic [. . .] but it is also *hetero*-genetic: the virtual generates or engenders the actual *out of* its intrinsic differentiality *through* differenciation. And it is because ontogenesis is heterogenesis, because every process of actualisation amounts to the production of *new* differences through which a

multiplicity is actualised, that Deleuze, following Bergson's conception of a creative evolution, characterises such processes as "creation" [...]' (de Beistegui, 2004: 293–4). I want to stress the parallels between this process account of genesis and Deleuze's process account of time. The genetic account given by de Beistegui is consistent with and explains the account in terms of time syntheses, for instance, in the correspondences between the new in both versions. The genetic account presupposes the philosophy of time and the syntheses of time are genetic.

3. For an interesting version of these questions around the human in Deleuze, see Hallward, 2006: 66–7. Hallward's argument depends on seeing Deleuze's move beyond the human as a move into the infinity of God (67). I do not see this as a necessary step. The danger in Deleuze is perhaps not in a move up to God, which is based on a misinterpretation of Deleuze's work on Spinoza and substance. Instead, the risk is that the comprehensiveness of his definition of individuation removes any tools for treating the human as more valuable in relation to other things.

4. These arguments are a response to points made about the distinction between times in *Difference and Repetition* and *Logic of Sense* by Jonathan Roffe in his unpublished PhD thesis *The Insubordinate Multiple: a Critique of Badiou's Deleuze* (Roffe, 2010: 134). Roffe gives an outstanding critique of Badiou's reading of Deleuze's philosophy of time in the thesis (127–52). In his scholarly and wide-ranging work on Deleuze's philosophy of time, Nathan Widder is perhaps the only commentator to have developed a reading of time that takes *Difference and Repetition* and *Logic of Sense* together but without operating a reduction on one or the other. Partly this comes from Widder's exceptional tracing of the influence of Freud and Lacan on Deleuze's philosophy of time, but even more so, it depends on his tenacious textual analysis, for instance in tracking down correspondences and divergences on Deleuze's two books between uses of the three syntheses: 'Deleuze here speaks of three syntheses, analogous to the three syntheses of time, that together form the field of sense: connective, conjunctive and disjunctive' (Widder, 2008: 111).

5. The concept of the simulacrum is crucial to understanding the relation between times and between the actual and virtual in Deleuze's thought. This is why I have reservations about approaches that leave it out, such as the study of the reciprocal determination of virtual and actual given by Miguel de Beistegui, despite the great originality of his reading of the concept of genesis in Deleuze. He approaches this determination with a Kantian and phenomenological vocabulary of noumenon and phenomenon which remains abstract in terms of actual processes and ties Deleuze too closely to phenomenology (de Beistegui, 2004: 227). This importation of terms is not necessary once the central role of simulacra is taken into account.

6. The most extensive study of this new kind of Deleuzian moral philosophy in a political context can be found in John Protevi's *Political Affect: Connecting the Social and the Somatic*. His analysis of the Terri Schiavo case on the grounds of a novel take on jurisprudence after Deleuze is a far-sighted and deeply humane take on the intersection of moral, emotional, religious, scientific and juridical in cases of the cessation of artificial feeding. The most original aspects of the analysis in the context of this study of Deleuze's philosophy of time in relation to moral problems are in Protevi's understanding that Deleuze's concept of singularity allows for a defence of privacy in conjunction with a conception of life as social: 'It is this intensity generated by concrete processes forming bodies politic that lies behind the justification of privacy as singularity' (Protevi, 2009: 190). The connected social sphere is a network of relation through affects and intensities that are singular in their actual individual manifestations. It is therefore not a paradox to insist on the social and privacy because neither the social nor the private allows for normalisation, because they depend on singular intensities for their connection (189). In terms of Deleuze's philosophy of time, the living present (private) is always a dimension of the pure past and of the future defined as eternal return of all differences (virtual intensities expressed in any part of the public and the political) and neither can be represented according to a normalising model reliable through time and across individuals.
7. I have given the non-mathematical definition of singularity here, in contrast for instance to DeLanda's interpretation (DeLanda, 2002: 127). This is because I want to insist, once again, on the independence of Deleuze's treatment of time in relation to problems and to multiplicities of singularities in order to demonstrate the metaphysical and speculative cohesiveness and flexibility of Deleuze's approach to time, such that it cannot be reduced to a mathematical model of time, however open and abstract it might be. For a very good account of this independence in the context of Deleuze's work on Lautman, see Domenech-Oneto and Roque, 2009: 176–7.
8. For a full discussion of Deleuze's philosophy of language as developed in *Logic of Sense*, see Lecercle, 2002; Williams, 2008a: 28–69.
9. In his comprehensive article on time and Deleuze, Peter Pál Pelbart describes the multiple times found in Deleuze's works as forming a 'mosaic'. This is a helpful image from an article that is very useful for tracing the great number and diversity of accounts of time in Deleuze. The article is also valuable in giving a version of time less easily traceable in Deleuze's earlier philosophy but found in later works such as *Foucault* and *The Fold: Leibniz and the Baroque*: time as pure exteriority (Pelbart, 1998: 101). A possible candidate for at least some kind of tracking back of this concept of time might be through the eternal return of pure difference.

7 CONCLUSION: THE PLACE OF FILM IN DELEUZE'S PHILOSOPHY OF TIME

1. There are very many works on Deleuze and film. D. N. Rodowick's *Gilles Deleuze's Time Machine* is an excellent source for work on Deleuze, time and film. See, in particular, his remarks on the fate of philosophy as connected to the fate of film and his work on the idea of a 'direct image' that is nonetheless not a representation: 'To say that time has its direct image does not necessarily mean that time is a form that can be represented' (Rodowick, 1997: 185). Rodowick's study stands as a counter to my remarks on the dangers of Deleuze's film books for his philosophy of time. This is partly because Rodowick adds to those books by reading them in relation to a much wider corpus and, perhaps most significantly, in relation to Nietzsche. See also Gregory Flaxman's important collection *The Brain is the Screen: Deleuze and the Philosophy of Cinema*. Flaxman's own article in the collection is a particularly useful account of Bergson's thought as it relates to Deleuze's work on cinema (Flaxman, 2000: 87). David Martin-Jones makes useful points with respect to the manner in which Deleuze goes beyond Bergson through Kant and Nietzsche in his *Deleuze, Cinema and National Identity: Narrative Time in National Contexts* (Martin-Jones, 2006: 60–1). For a characteristically original and philosophically inventive work on Deleuze and cinema in the French tradition, see Pierre Montebello's *Deleuze, philosophie et cinéma*. Montebello's work is particularly deep because he makes appeal to the concept of paradox with respect to time that he also reflects upon in his other books, in order to conclude with 'seven paradoxes on movement and time' where he uses the beautiful phrase that Bergson's theses are raised to the 'incandescence of the paradox' by Deleuze (Montebello, 2008b: 131). In writing this book I also benefited greatly from being able to read Patricia Pisters's latest work on film prior to publication; for her earlier work on Deleuze and film, see Pisters, 2003. Last but not least, my decision to leave out Deleuze's work on film from the main body of my work on time was strongly influenced by the scholarly and acute critique of Deleuze's reading of Bergson in the cinema books. In particular, Mullarkey's understanding of the dangers of the use of the term 'image' in relation to film and to Bergson, and his careful critique of the movement-image and time-image distinction, were essential to me (Mullarkey, 2009: 97–100).
2. See Montebello, 2008b; Rodowick, 1997; Mullarkey, 2009 for examples of this kind of critical and creative extension of Deleuze's work on film. Such works can rightly be seen not as work on Deleuze's film theory, but rather as works after his film-philosophy.
3. See Mullarkey, 2009: 100–3.
4. Perhaps the best work for understanding the explanatory scope and interpretative promise of Deleuze's works for the study and indeed

aesthetic appreciation of film is Ronald Bogue's *Deleuze on Cinema*. See, in particular, Bogue's study of the crystal image where the exemplification, for instance through an analysis of Garbo, is a model of clarity and depth with the extra merit of insisting on the genetic aspect of the crystal, something that is often overlooked, yet nonetheless crucial to understanding the relation to the future of the image (Bogue, 2003: 123).

5. See Widder, 2008: 90 for a strong critical explanation of the extension and drift from Bergson allowed by Deleuze's reading of Nietzsche. For a counter-view, that Widder is reacting to, see Moulard, 2002, where Valentine Moulard argues that Bergson's work allows Deleuze to move beyond a Kantian transcendental frame in his cinema books.

6. Pierre Montebello makes similar points but with much greater emphasis on the correspondences through the paradox to Deleuze's other works: 'every event implies simultaneity of the presents' . . . 'another form of the interiority of time' (Montebello, 2008b: 125–6). My view is that we have to make qualitative distinctions about the nature of paradoxes in Deleuze around their genetic power. The problem is that the paradoxes around simultaneity and interiority are generated by discordances with commonplace perception and standard logic. In turn this generates overly simple, that is, analogical solutions drawn from film, a kind of reflection driven by the thought that 'in some way things are thus even though they are not really so'. The paradoxes from *Difference and Repetition* and *Logic of Sense* lead to the thought that the real is generated by paradoxes and is inherently paradoxical. This demands a constant creation of the real in response to its paradoxical nature, where analogical models already elide the paradox they are meant to work with.

Bibliography

NOTE

The excellent research source, webdeleuze.com, has a superb Deleuze bibliography by Timothy S. Murphy, 'Revised Bibliography of the Works of Gilles Deleuze', at http://www.webdeleuze.com.

WORKS BY GILLES DELEUZE

Empirisme et subjectivité: Essai sur la Nature humaine selon Hume (Paris: Presses universitaires de France, 1953). Boundas, C., *Empiricism and Subjectivity: An Essay on Hume's Theory of Human Nature* (New York: Columbia University Press, 1991).

'Cours sur le chapitre III de L'évolution créatrice' in Worms, F. (ed.), *Annales bergsoniennes: Tome 2, Bergson, Deleuze, la phénoménologie* (Paris: Presses universitaires de France, 1960).

Nietzsche et la philosophie (Paris: Presses universitaires de France, 1962). Trans. Tomlinson, H., *Nietzsche and Philosophy* (New York: Columbia University Press, 1983).

La Philosophie critique de Kant: Doctrine des facultés (Paris: Presses universitaires de France, 1963). Trans. Tomlinson, H. and Habberjam, B., *Kant's Critical Philosophy: The Doctrine of the Faculties* (Minneapolis, MN: University of Minnesota Press, 1984).

Le Bergsonisme (Paris: Presses universitaires de France, 1966). Trans. Tomlinson, H. and Habberjam, B., *Bergsonism* (New York: Zone Books, 1990).

'Gilbert Simondon. – *L'Individu et sa genèse physico-biologique*' (book review), in *Revue philosophique de la France et de l'étranger*, CLVI:1–3 (janvier–mars 1966), pp. 115–18. Trans. Ramirez, I., 'Review of Gilbert Simondon's *L'Individu et sa genèse physico-biologique* (1966)', *Pli, The Warwick Journal of Philosophy*, Vol. 12 (2001), pp. 43–9.

Bibliography

Présentation de Sacher-Masoch (Paris: Éditions de Minuit, 1967). Trans. McNeil, J., *Masochism* (New York: Zone Books, 1989).

Différence et répétition (Paris: Presses universitaires de France, 1968). Trans. Patton, P., *Difference and Repetition* (New York: Columbia University Press, 1994).

Spinoza et le problème de l'expression (Paris: Éditions de Minuit, 1968). Trans. Joughin, M., *Expressionism in Philosophy: Spinoza* (New York: Zone Books, 1990).

Logique du sens (Paris: Éditions de Minuit, 1969). Trans. Lester, M. and Stivale, C., *The Logic of Sense* (New York: Columbia University Press, 1990).

Proust et les signes (Paris: Presses universitaires de France, 1970). Trans. Howard, R., *Proust and Signs* (New York: George Braziller, 1972).

and Félix Guattari, *Capitalisme et schizophrénie tome 1: l'Anti-Oedipe* (Paris: Éditions de Minuit, 1972). Trans. Hurley, R., Seem, M. and Lane, H., *Anti-Oedipus: Capitalism and Schizophrenia* (New York: Viking Press, 1977).

and Félix Guattari, *Kafka: Pour une littérature mineure* (Paris: Éditions de Minuit, 1975). Trans. Polan, D., *Kafka: Toward a Minor Literature* (Minneapolis, MN: University of Minnesota Press, 1986).

and Claire Parnet, *Dialogues* (Paris: Flammarion, 1977). Trans. Tomlinson, H. and Habberjam, B., *Dialogues* (New York: Columbia University Press, 1987).

and Félix Guattari, *Capitalisme et schizophrénie tome 2: Mille plateaux* (Paris: Éditions de Minuit, 1980). Trans. Massumi, B., *A Thousand Plateaus: Capitalism and Schizophrenia* (Minneapolis, MN: University of Minnesota Press, 1987).

Spinoza: Philosophie pratique (Paris: Editions de Minuit, 1981). Trans. Hurley, R., *Spinoza: Practical Philosophy* (San Francisco: City Lights, 1988).

Francis Bacon: Logique de la Sensation (Paris: Éditions de la Différence, 1981). Trans. Smith, D., *Francis Bacon: the Logic of Sensation* (Minneapolis, MN: University of Minnesota Press, 2003).

Cinéma-1: L'Image-mouvement (Paris: Éditions de Minuit, 1983). Trans. Tomlinson, H. and Habberjam, B., *Cinema 1: The Movement-Image* (Minneapolis, MN: University of Minnesota Press, 1986).

Cinéma-2: L'Image-temps (Paris: Éditions de Minuit, 1985). Trans. Tomlinson, H. and Galeta, R., *Cinema 2: The Time-Image* (Minneapolis, MN: University of Minnesota Press, 1989).

Foucault (Paris: Éditions de Minuit, 1986). Trans. Hand, S., *Foucault* (Minneapolis, MN: University of Minnesota Press, 1988).

Le Pli: Leibniz et le Baroque (Paris: Éditions de Minuit, 1988). Trans. Conley, T., *The Fold: Leibniz and the Baroque* (Minneapolis, MN: University of Minnesota Press, 1993).

Pourparlers 1972–1990 (Paris: Éditions de Minuit, 1990). Trans. Joughin, M., *Negotiations 1972–1990* (New York: Columbia University Press, 1995).

and Félix Guattari, *Qu'est-ce que la philosophie?* (Paris: Éditions de Minuit, 1991). Trans. Tomlinson, H. and Burchell, G., *What is Philosophy?* (New York: Columbia University Press, 1994).

'L'épuisé', in Samuel Beckett, *Quad et trio du fantôme*, trans. E. Fournier (Paris: Éditions de Minuit, 1992), pp. 57–106.

Boundas, Constantin V. (ed.), *The Deleuze Reader* (New York: Columbia University Press, 1993).

Critique et clinique (Paris: Editions de Minuit, 1993). Trans. Smith, D. and Greco, A., *Essays Critical and Clinical* (Minneapolis, MN: University of Minnesota Press, 1997).

et al., 'Gilles Deleuze', *Philosophie*, no. 47 (1995) (includes the important last essay by Deleuze, 'L'Immanence: une vie . . .').

L'Île déserte et autres textes. Textes et entretients 1953–1974 (Paris: Éditions de Minuit, 2002). Trans. Taormina, M., *Desert Islands and Other Texts* (New York: Semiotext(e), 2003).

Deux régimes de fous. Textes et entretients 1975–1995 (Paris: Éditions de Minuit, 2003).

OTHER WORKS

Agamben, Giorgio, *Potentialities: Collected Essays in Philosophy* (Palo Alto, CA: Stanford University Press, 2000).

Alliez, Éric, *Capital Times: Tales from the Conquest of Time*, trans. G. Van Den Abeele (Minneapolis, MN: University of Minnesota Press, 1996).

Ansell Pearson, Keith, *Germinal Life: the Difference and Repetition of Gilles Deleuze* (London: Routledge, 1999).

Ansell Pearson, Keith, *Philosophy and the Adventure of the Virtual: Bergson and the Time of Life* (London: Routledge, 2002).

Aristotle, *Physics*, trans. R. Waterfield (Oxford: Oxford University Press, 2008).

Armstrong, Michael, *Managing People: A Practical Guide for Line Managers* (London: Kogan Page, 1998).

Armstrong, Michael, *A Handbook of Human Resource Management Practice* (London: Kogan Page, 2006).

Augustine, *Confessions*, trans. H. Chadwick (Oxford: Oxford University Press, 1998).

Badiou, Alain, *Deleuze: la clameur de l'être* (Paris: Hachette, 1997).

Badiou, Alain, *Theoretical Writings*, ed. R. Brassier and A. Toscano (London: Continuum, 2004).

Battersby, Christine, *The Phenomenal Woman: Feminist Metaphysics and the Patterns of Identity* (Cambridge: Polity, 1998).

Beistegui, Miguel de, *Truth and Genesis: Philosophy as Differential Ontology* (Bloomington, IN: Indiana University Press, 2004).

Bell, Jeffrey, *Philosophy at the Edge of Chaos: Gilles Deleuze and the Philosophy of Difference* (Toronto: University of Toronto Press, 2006).

Bibliography

Bell, Jeffrey, *Deleuze's Hume: Philosophy, Culture and the Scottish Enlightenment* (Edinburgh: Edinburgh University Press, 2009).
Bell, Martin, 'Transcendental empiricism? Deleuze's reading of Hume', in M. Frasca Spada (ed.), *Impressions of Hume* (Oxford: Oxford University Press, 2005).
Bergson, Henri, *Oeuvres* (Édition du centenaire) (Paris: Presses universitaires de France, 1959).
Bogue, Ronald, *Deleuze on Cinema* (New York: Routledge, 2003).
Boundas, Constantin, *Deleuze and Philosophy* (Edinburgh: Edinburgh University Press, 2006).
Bryant, Levi R., *Difference and Givenness: Deleuze's Transcendental Empiricism and the Ontology of Immanence* (Evanston, IL: Northwestern University Press, 2008).
Buchanan, Ian, 'Is Anti-Oedipus a May '68 book?', in J. A. Bell and C. Colebrook (eds), *Deleuze and History* (Edinburgh: Edinburgh University Press, 2009), pp. 206–24.
Cazenave, Catherine, 'Le coup de dès ou l'affirmation du hasard', in *Cahiers Critiques de Philosophie*, no. 2 (avril 2006), pp. 103–16.
Colebrook, Claire, 'Introduction', in J. A. Bell and C. Colebrook (eds), *Deleuze and History* (Edinburgh: Edinburgh University Press, 2009), pp. 1–32.
Conley, Tom, 'From image to event: reading Genet through Deleuze', in *Yale French Studies*, no. 91 (1997), pp. 49–63.
DeLanda, Manuel, *Intensive Science and Virtual Philosophy* (London: Continuum, 2002).
Derrida, Jacques, *Spectres of Marx: the State of Debt, the Work of Mourning, and the New International*, trans. P. Kamuf (London: Routledge, 1994).
Domenech-Oneto, Paulo and Roque, Tatiana (2009), 'L'objectivité des problèmes et la question du sujet: considerations sur l'Idée dans la philosophie de Deleuze', in P. Cassou-Noguès and P. Gillot (eds), *Le concept, le sujet et la science* (Paris: Vrin, 2009), pp. 165–90.
Faulkner, Keith, *The Force of Time: an Introduction to Deleuze through Proust* (Lanham, MD: University Press of America, 2008).
Flaxman, Gregory (ed.), *The Brain is the Screen: Deleuze and the Philosophy of Cinema* (Minneapolis, MD: University of Minnesota Press, 2000).
Frichot, Hélène, 'The Onanist's escape from architectural captivity', at http://www.dab.uts.edu.au/conferences/queer_space/proceedings/architecture_frichot.pdf (2007; accessed 19 May 2010).
Grosz, Elizabeth, *The Nick of Time: Politics, Evolution and the Untimely* (Durham, NC: Duke University Press, 2004).
Hallward, Peter, *Out of this World: Deleuze and the Philosophy of Creation* (London: Verso, 2006).
Hawking, Stephen, *A Brief History of Time: from the Big Bang to Black Holes* (London: Bantam, 1988).

Heidegger, Martin, *Being and Time*, trans. J. Macquarrie and E. Robinson (Oxford: Blackwell, 1985).
Heidegger, Martin, *Nietzsche, Volumes Three and Four*, ed. D. F. Krell (San Francisco: HarperCollins, 1991).
Hölderlin, Friedrich, *Essays and Letters on Theory*, trans. T. Pfau (New York: SUNY Press, 1988).
Holland, Eugene W., *Deleuze and Guattari's Anti-Oedipus: Introduction to Schizoanalysis* (London: Routledge, 1999).
Hume, David, *A Treatise of Human Nature*, ed. D. Norton and M. Norton (Oxford: Oxford University Press, 2009).
Jones, Graham and Roffe, Jon, *Deleuze's Philosophical Lineage* (Edinburgh: Edinburgh University Press, 2009).
Kant, Immanuel, *Critique of Pure Reason*, trans. N. Kemp Smith (London: Macmillan, 1985).
Kerslake, Christian, *Immanence and the Vertigo of Philosophy* (Edinburgh: Edinburgh University Press, 2009).
Khalfa, Jean, *An Introduction to the Philosophy of Gilles Deleuze* (London: Continuum, 2003).
Kierkegaard, Søren, *Concluding Unscientific Postscript* to Philosophical Fragments, ed. and trans. H. V. Hong and E. H. Hong (Princeton, NJ: Princeton University Press, 1992).
Klossowski, Pierre, *Nietzsche and the Vicious Circle*, trans. D. Smith (London: Athlone, 1997).
Lampert, Jay, *Deleuze and Guattari's Philosophy of History* (London: Continuum, 2006).
Lampert, Jay 'Theory of delay in Balibar, Freud and Deleuze', in J. A. Bell and C. Colebrook (eds), *Deleuze and History* (Edinburgh: Edinburgh University Press, 2009), pp. 72–91.
Lawlor, Leonard, 'The beginnings of thought: the fundamental experience in Derrida and Deleuze', in P. Patton and J. Protevi (eds), *Between Deleuze and Derrida* (London: Continuum, 2003), pp. 67–83.
Lecercle, Jean-Jacques, *Deleuze and Language* (Basingstoke: Palgrave Macmillan, 2002).
Leibniz, G. W., *Philosophical Texts*, ed. and trans. R. S. Woolhouse and R. Francks (Oxford University Press, 1998).
Le Poidevin, Robin, *Travels in Four Dimensions: the Enigmas in Space and Time* (Oxford: Oxford University Press, 2003).
Lewis, David, 'The paradoxes of time travel', in *Philosophical Papers Volume II* (Oxford: Oxford University Press, 1986).
Lockwood, Michael, *The Labyrinths of Time: Introducing the Universe* (Oxford: Oxford University Press, 2005).
Lorraine, Tamsin, 'Living a time out of joint', in P. Patton and J. Protevi (eds), *Between Deleuze and Derrida* (London: Continuum, 2003), pp. 30–44.

Bibliography

Lyotard, Jean-François, *The Differend*, trans. G. Van Den Abeele (Manchester: University of Manchester Press, 1988).

Lyotard, Jean-François, *La confession d'Augustin* (Paris: Galilée, 1998).

McTaggart, J. M. E., 'The unreality of time', in R. Le Poidevin and M. MacBeath (eds), *The Philosophy of Time* (Oxford: Oxford University Press, 1993), pp. 23–34.

Martin-Jones, David, *Deleuze, Cinema and National Identity: Narrative Time in National Contexts* (Edinburgh: Edinburgh University Press, 2006).

Massumi, Brian, *Parables for the Virtual: Movement, Affect, Sensation* (Durham, NC: Duke University Press, 2002).

Meillassoux, Quentin, *Après la finitude: essai sur la nécessité de la contingence* (Paris: Seuil, 2006).

Merleau-Ponty, Maurice, *Phénoménologie de la perception* (Paris: Gallimard, 1945).

Mill, John Stuart, *Utilitarianism* (Indianapolis, IN: Hackett, 2002).

Montebello, Pierre, *Deleuze* (Paris: Vrin, 2008).

Montebello, Pierre, *Deleuze, philosophie et cinéma* (Paris: Vrin, 2008).

Moulard, Valentine, 'The time-image and Deleuze's transcendental experience', in *Continental Philosophy Review*, vol. 35, no. 3 (2002), pp. 325–45.

Mullarkey, John, *Post-Continental Philosophy: an Outline* (London: Continuum, 2006).

Mullarkey, John, *Refractions of Reality: Philosophy and the Moving Image* (Basingstoke: Palgrave Macmillan, 2009).

Nietzsche, Friedrich, *Thus Spoke Zarathustra*, trans. R. J. Hollingdale (London: Penguin, 1969).

Olkowski, Dorothea, *Gilles Deleuze and the Ruin of Representation* (Berkeley, CA: University of California Press, 1999).

Olkowski, Dorothea, *The Universal (in the Realm of the Sensible): Beyond Continental Philosophy* (Edinburgh: Edinburgh University Press, 2007).

Olkowski, Dorothea, 'Deleuze and the limits of mathematical time', in *Deleuze Studies*, vol. 2, no. 1 (2008), pp. 1–24.

Panagia, Davide, 'Inconstancies of character: David Hume on sympathy, intensity and artifice', in Boundas (2006), pp. 85–97.

Patton, Paul, 'Events, becoming and history', in J. A. Bell and C. Colebrook (eds), *Deleuze and History* (Edinburgh: Edinburgh University Press, 2009), pp. 33–53.

Péguy, Charles, *Clio* (Paris: Gallimard, 1932).

Pelbart, Peter Pál, 'Le temps non-réconcilié', in E. Alliez (ed.), *Gilles Deleuze: une Vie Philosophique* (Paris: Les empêcheurs de penser en rond, 1998), pp. 89–102.

Pisters, Patricia, *The Matrix of Visual Culture: Working with Deleuze in Film Theory* (Palo Alto, CA: Stanford University Press, 2003).

Pisters, Patricia, *The Neuro-Image: A Deleuzian Filmphilosophy for Digital Screen Culture* (Palo Alto, CA: Stanford University Press, forthcoming).

Plato, *Symposium*, ed. K. J. Dover (Cambridge: Cambridge University Press, 1980).
Plato, *Complete Works*, ed. J. M. Cooper (London: Hackett, 1997).
Prigogine, Ilya and Stengers, Isabelle, *Entre le temps et l'éternité* (Paris: Flammarion, [1988] 1992).
Protevi, John, *Political Affect: Connecting the Social and the Somatic* (Minneapolis, MN: University of Minnesota Press, 2009).
Reichenbach, Hans, *The Philosophy of Space and Time*, trans. M. Reichenbach and J. Freund (New York: Dover, 1958).
Reynolds, Jack, 'Wounds and scars: Deleuze on the time and ethics of the event', in *Deleuze Studies*, vol. 1, no. 2 (2007), pp. 144–66.
Reynolds, Jack, 'Transcendental priority and Deleuzian normativity: a reply to James Williams', in *Deleuze Studies*, vol. 2, no. 1 (2008), pp. 101–8.
Reynolds, Jack and Roffe, Jonathan, 'Deleuze and Merleau-Ponty: Immanence, Univocity, and Phenomenology', in *The Journal of the British Society of Phenomenology*, vol. 37, no. 3 (October 2006), pp. 228–51.
Ricoeur, Paul, *Time and Narrative* (Vol. 1), trans. K. McLaughlin and D. Pellauer (Chicago, IL: University of Chicago Press, 1984).
Rodowick, D. N., *Gilles Deleuze's Time Machine* (Durham, NC: Duke University Press, 1997).
Roffe, Jonathan, 'David Hume', in Jones and Roffe (2009), pp. 67–86.
Roffe, Jonathan, *The Insubordinate Multiple: a Critique of Badiou's Deleuze* (unpublished PhD thesis, University of Tasmania, 2010).
Sartre, Jean-Paul, *L'être et le néant: essai d'ontologie phénoménologique* (Paris: Gallimard, 1943).
Sartre, Jean-Paul, *Saint Genet: comédien et martyr* (Paris: Gallimard, 1952).
Savitt, Steven F., *Time's Arrows Today: Recent Physical and Philosophical Work on the Direction of Time* (Cambridge: Cambridge University Press, 1995).
Schmitt, Frederick, 'Why was Descartes a foundationalist?', in A. Rorty (ed.), *Essays on Descartes' Meditations* (Berkeley, CA: University of California Press, 1986).
Shakespeare, William, *Hamlet* (London: Penguin, 2007).
Shoemaker, Sydney, 'Time without change', in R. Le Poidevin and M. MacBeath (eds), *The Philosophy of Time* (Oxford: Oxford University Press, 1993), pp. 63–79.
Simont, Juliette, 'Intensity, or: the "encounter"', in Khalfa (2003), pp. 26–49.
Smith, Daniel W., 'The concept of the simulacrum: Deleuze and the overturning of Platonism', in *Continental Philosophy Review*, vol. 38, nos 1–2 (2006), pp. 89–113.
Smith, Daniel W., 'The conditions of the new', in *Deleuze Studies*, vol. 1, no. 1 (2007), pp. 1–21.
Smith, Daniel W., 'G. W. F. Leibniz', in Jones and Roffe (2009), pp. 44–66.
Turetzky, Philip, *Time* (London: Routledge, 1998).

Bibliography

Waterlow, Sarah, *Nature, Change and Agency in Aristotle's Physics* (Oxford: Oxford University Press, 1982).

Waterlow, Sarah, 'Aristotle's now', in *The Philosophical Quarterly*, vol. 34, no. 135 (1984), pp. 104–28.

Widder, Nathan, *Reflections on Time and Politics* (University Park, PA: Pennsylvania State University Press, 2008).

Williams, James, 'Narrative and time', in *Proceedings of the Aristotelian Society*, Supplementary Volume (Spring 1996), pp. 47–61.

Williams, James, *Gilles Deleuze's* Logic of Sense: *A Critical Introduction and Guide* (Edinburgh: Edinburgh University Press, 2008a).

Williams, James, 'Why Deleuze does not blow the actual on virtual priority. A rejoinder to Jack Reynolds', in *Deleuze Studies*, vol. 2, no. 1 (2008b), pp. 97–100.

Williams, James, 'Ageing, perpetual perishing and the event as pure novelty: Péguy, Whitehead and Deleuze on time and history', in J. A. Bell and C. Colebrook (eds), *Deleuze and History* (Edinburgh: Edinburgh University Press, 2009).

Williams, James, 'Against oblivion and simple empiricism: Gilles Deleuze's "Immanence: a life . . .",' in *Journal of Philosophy: a Cross-Disciplinary Inquiry*, vol. 5, no. 10 (Winter 2010), pp. 35–44.

Wood, David, *Time after Time* (Bloomington, IN: Indiana University Press, 2007).

Worms, Frédéric, (2004) 'La conscience ou la vie? Bergson entre phénoménologie et métaphysique' in F. Worms (ed.), *Annales bergsoniennes: Tome 2, Bergson, Deleuze, la phénoménologie* (Paris: Presses Universitaires de France, 2004), pp. 191–206.

Index

Actual, 8, 10, 13–14, 16–17, 19, 26, 28–34, 42, 45–6, 57, 60–4, 71–7, 84, 108, 111, 128, 136, 140–5, 147–57
Agamben, 178n
Aiôn, 9, 138–40, 145–7, 150, 152–8, 163
Alliez, 175–6n
Analogy, 98
Ansell Pearson, 166n, 172n, 180n, 184n, 185–6n
Aristotle, 167n, 174–5n, 179n
Armstrong, 165n
Arrow of time, 4, 28, 30, 34, 46, 59, 167n
Asymmetry, 2–6, 28, 59, 98, 100–1, 122, 152, 164
Augustine, 46, 159, 163, 166n, 170n, 176n

Badiou, 173–4n, 185n, 186n, 189–90n
Bartleby, 102
Battersby, 185n
Becoming, 2–3, 4, 13–14, 18–19, 37, 40, 42, 48–9, 65, 67, 103, 123–4, 126, 135, 138–57
Beistegui, de, 179n, 189–90n
Bell, J., 165n, 166n, 167n, 176–7n, 188n
Bell, M., 169n

Bergson, 1, 12, 34, 46, 63, 66–9, 73, 77, 84, 93, 110, 159–63, 167n, 168n, 169n, 170n, 171n, 176–7n, 177n, 180n, 188–9n, 192n, 193n
Bogue, 193n
Bryant, 171–2n, 180n, 183n, 184n

Caesura, 14–15, 89–91, 99–101, 114
Carroll, 131, 140, 160
Cause, 38–9, 41, 59, 148–52
Cazenave, 186n
Chance, 125–6
Chronos, 9, 138, 140, 145–7, 151–8, 163
Cinema, 2, 159–64
Conley, 180n
Counter-actualisation, 145, 153

Dark precursor, 87, 131–3, 145
Darwin, 178n
Death, 118–23
DeLanda, 179n, 191n
Derrida, 181–2n
Descartes, 79–83, 176n, 180n
Destiny, 57, 66–72, 146–8, 158, 161
Determinism, 70
Difference (pure), 13–16, 18, 20, 96, 98, 101, 106, 109, 111, 115–19, 121–3, 126, 128, 130–1, 135–6, 152–3

Index

Dimensions of time, 5–6, 9, 65–6, 69, 94, 103–4, 113, 127, 138, 154, 163
Disjunction, 45, 151–2, 162
Dogmatism, 106–12
Domenech-Oneto, 191n
Dramatisation, 13–14, 87–8, 90–1, 93–4, 97, 108

Einstein, 168n
Empedocles, 102
Eros, 78–9, 109, 111–12, 114
Eternal return, 12, 16, 79, 84–5, 109, 111–12, 113ff, 136–7, 141, 154, 162
Event, 5–16, 18–19, 25–8, 29–31, 33–4, 43–8, 72–3, 88–9, 91–101, 106–9, 116–19, 125–8, 135–7, 139, 143–5, 149, 152–8
Existentialism, 119–21

Faulkner, 171n
First synthesis of time, 10–12, 17–18, 25ff, 51–2, 55, 65, 69, 95, 103, 113, 120, 135, 146–7
Flaxman, 192n
Foundation, 2, 14, 25, 52, 55–6, 76, 80, 83–9, 93, 103–4, 162, 175n
Founding, 55–7, 85–6, 103–4, 162
Freedom, 57, 71–3
Freud, 109–10, 123, 190n
Frichot, 180n
Future *see* eternal return; new; first, second and third syntheses of time

Genesis, 1, 26, 161, 183n
Genet, 74, 180n
Grosz, 183–4n
Guattari, 2, 173n, 175–6n, 180n, 186n

Habit, 37, 38–45, 50, 59, 95–6, 113, 116, 120
Hallward, 171n, 177–8n, 190n
Hamlet, 87–9, 99, 102, 136, 181–2n
Hawking, 167n
Heidegger, 1, 185n

Heraclitus, 130
History, 19–20, 31, 95–102
Hölderlin, 83–4, 87, 90, 136, 180–1n, 182n
Holland, 172–3n
Hume, 12, 21, 24, 27, 30, 33–4, 37, 42, 46, 60, 165n, 166n, 169n, 170n, 176–7n

Image, 92–3, 95–102, 104, 108, 160–3
Individuation, 32, 42, 44, 137–8, 145, 152
Intensity, 18, 40–1, 72, 138, 140, 143–5, 147, 153–4, 160

Joyce, 131

Kant, 1, 12, 73, 79–86, 168n, 171n, 180n, 183n, 184n, 190n, 192n, 193n
Kauffman, 167n
Kierkegaard, 185n
Klein, 110
Klossowski, 184–5n
Knowledge, 12, 84–5, 132, 142, 144, 152

Lacan, 110, 190n
Lampert, 172n, 173n, 176–6n, 177n, 188n
Language, 122–3, 144, 146–51, 161
Larval subjects, 49–50, 87, 145
Lautman, 191n
Lawlor, 182n
Le Poidevin, 174–5n
Learning, 37–8
Lecercle, 191n
Leibniz, 23, 119, 165n, 166n, 170n, 175n, 180n, 187n, 191n
Lewis, 170n, 179n, 182n
Living present, 14–15, 21–30, 33, 35–6, 41, 47, 60, 74, 77, 105, 113, 116–17, 121, 127, 129, 135, 137, 140, 146–52
Lockwood, 167n

203

Lorraine, 181–2n
Lyotard, 175–6n

Maldiney, 110
Mallarmé, 186n
Martin-Jones, 192n
Marx, 178n
Massumi, 189n
Meillassoux, 168n
Memory, 33, 51, 54–5, 57, 59, 61–2, 68, 76, 85, 106, 120, 135–6
Merleau-Ponty, 30, 168n
Metaphysics, 7, 42, 54–5, 73, 120, 160
Method, 2, 7, 30, 32–6, 40, 69, 80, 87, 89, 97, 100, 107, 129; *see also* process; speculative philosophy
Mill, 74
Montebello, 177n, 186n, 192n, 193n
Moral principles, 76–7, 143–4, 154–8
Moulard, 193n
Mullarkey, 171n, 192n
Multiplicity, 3–6, 13, 44, 70, 94, 137, 164

New, 93, 109, 116
Nietzsche, 12, 16, 79, 84–5, 111, 114, 118–19, 136, 154, 162, 170n, 183–4n, 185n, 186n, 188–9n, 192n

Object, 17, 32–3, 110–11, 129, 141, 144
Oedipus, 102, 136
Olkowski, 167n, 168n, 188–9n

Pál Pelbart, 191n
Panagia, 165n
Paradox, 22, 45–6, 52–3, 63–4, 83, 130, 142–3, 170n
Passive synthesis, 21, 25–6, 44, 81
Past *see* first, second and third syntheses of time; pure past
Patton, P., 27, 32, 44
Péguy, 175n
Phenomenology, 31, 47
Physics, 1, 3, 14, 47, 114
Pisters, 192n

Plato, 1, 42, 84–6, 88, 101, 114, 138, 141–3, 181n, 187n, 188–9n
Possibility, 26, 28, 30
Present *see* first, second and third syntheses of time; living present
Prigogine, 167n
Process, 2–3, 8, 30, 54, 65–6, 104, 113, 115, 117, 124, 128, 134
Protevi, 191n
Proust, 37, 77–8, 131, 171n
Psychology, 29, 47, 93
Pure past, 14–15, 63–6, 71–2, 74, 77, 86, 103–4, 111, 125, 129–30, 136–7, 150, 153

Reciprocal determination, 34, 53, 73–4, 128–54, 163
Reichenbach, 168n
Repetition, 22–4, 49, 55, 68, 75, 95–7, 106–7, 115
Representation, 13, 61, 77, 86, 95–6, 98, 100, 116–17, 122, 129, 161
Reynolds, 168n, 183n
Ricoeur, 183n
Riderhood, 102
Rodowick, 192n
Roffe, 165n, 168n, 190n
Roque, 191n
Roussel, 131

Sartre, 74, 180n
Savitt, 167n
Schmitt, 176n
Science, 2, 39, 41–2, 47, 107–8, 110, 114, 117
Second synthesis of time, 10–12, 17–18, 51ff, 84, 95, 103–5, 113, 120, 125, 135, 142
Series, 124–31
Shakespeare, 89, 94, 136
Shoemaker, 179n
Sign, 37–8, 45, 48–50, 57, 76, 169n
Simont, 166n
Simulacra, 127–32, 141–5
Singularity, 2–5, 50, 135, 145

Index

Smith, 166n, 170n, 175n, 181n, 187n, 188–9n
Sophocles, 83, 87, 180–1n
Soul, 42–3, 69, 73–4
Speculative philosophy, 2, 8, 30–6, 42, 45–7, 53–6, 63, 68, 73, 76, 87, 90–1, 100, 107, 114, 120, 129, 131, 135
Spinoza, 100–1, 153, 177n, 190n
Stengers, 167n
Stoics, 1, 9, 146
Subject, 16–17, 32–3, 79–82, 87, 90, 94, 97, 102, 104, 144, 146–7
Synthesis *see* first, second and third syntheses of time

Third synthesis of time, 10–12, 79ff, 113, 116, 121–2
Time (irreversible), 4, 59, 98, 122, 135; *see also* arrow of time
Time management, 17–18
Time (multiplicity of), 4, 11, 159, 176n
Time travel, 7–16
Transcendental, 30, 34–5, 58, 62, 68, 72–5, 80, 84, 114, 129, 135
Truth, 67, 108, 114, 135, 173n
Turetzky, 167n

Virtual, 8, 74, 111, 162

Waterlow, 167n
Whitehead, 2, 165n
Widder, 185n, 190n, 193n
Williams, 165n, 175n, 178n, 183n, 191n
Wood, 176n
Worms, 171n

Zola, 185–6n

EU representative:
Easy Access System Europe
Mustamäe tee 50, 10621 Tallinn, Estonia
Gpsr.requests@easproject.com

www.ingramcontent.com/pod-product-compliance
Lightning Source LLC
Chambersburg PA
CBHW051058230426
43667CB00013B/2355